AF808585

CBEST Math Practice Tests
Expanded Edition

Math Study Guide for CBEST Test Preparation for the California Basic Educational Skills Test

CBEST Math Practice Tests Expanded Edition: Math Study Guide for CBEST Test Preparation for the California Basic Educational Skills Test

ISBN: 978-1-949282-74-0

NOTE: The drawings in this publication are for illustration purposes only. They are not drawn to an exact scale.

For information on bulk discounts, please contact Exam SAM via our webpage.

TABLE OF CONTENTS

Format of the CBEST Mathematics Test

Three types of skills are assessed on the CBEST test of mathematics:

- Skill 1: The first skill group assessed on the exam is estimation, measurement, and basic statistical principles.
    - Estimation problems involve rounding figures up or down and then performing addition, subtraction, multiplication, or division.
    - Estimation problems also involve making estimates of time to complete a task or a journey.
    - Measurement problems cover knowledge of dimensions or length, weight, and capacity.
    - Measurement problems on the CBEST often include diagrams.
    - Statistical problems on the exam are normally related to the calculation of test scores, although there may be other types of statistical questions.
- Skill 2: The second skill group covered on the CBEST math test is computation and problem solving.
    - Computations will involve addition, subtraction, multiplication, and division.

    - Problem solving questions will normally present a practical problem for you to work out, such as calculating the amount of discount on an item on sale.
- Skill 3: The third skill group on the test is numerical and graphic relationships.
    - Numerical relationship questions will ask you to determine whether a given number is less than or greater than other numbers.
    - For graphic relationship questions, you will usually have to interpret data from a table, chart, or graph.

Types of Questions on the CBEST Math Test

Estimation and Measurement

The CBEST exam includes these types of estimation and measurement questions:

- Understanding how to measure temperature, length, weight, and capacity using United States' measurement systems
- Measuring linear distances and perimeter
- Estimating the time required in order to achieve a work-related objective
- Estimating the results of problems involving addition, subtraction, multiplication, and division without doing full computations

Statistical Principles

Statistical questions on the CBEST cover the following skills:

- Performing arithmetic on basic data related to test scores, such as averages, ratios, and percentages
- Interpreting standardized test scores such as stanine scores and percentiles
- Using test scores to understand how a particular student has performed relative to other students

- Understanding basic probability to make predictions based on the data provided

Computation and Problem Solving

Computation and problem solving questions will cover these skills:

- Adding, subtracting, multiplying, and dividing
- Performing calculations on whole numbers, both positive and negative
- Performing calculations on fractions, decimals, and percentages
- Solving practical problems, such as determining prices per unit
- Solving algebraic equations with one unknown variable
- Determining if enough information is provided in order to solve a problem
- Identifying the facts given in a problem
- Understanding alternative methods for solving problems

Numerical and Graphic Relationships

The exam includes these types of numerical and graphic relationship questions:

- Recognizing relationships in data, such as calculating a percentage increase or decrease
- Ordering fractions from greatest to least or least to greatest
- Determining if a given number is less than or greater than other numbers
- Using less than, greater than, and equal to express mathematical relationships
- Identifying mathematical equivalents, such as $^{1}/_{5}$ equals 20%
- Using rounding to solve problems
- Understanding word problems that contain logical relationships, such as "if-then" sentences or quantifiers like "some" or "no one."
- Identifying data that is missing from a table or graph
- Using data in tables, graphs, or charts to solve problems

How to Use This Study Guide

The practice math tests in this study guide contain questions of all of the types that you will see on the real CBEST test.

Practice test 1 in this book is in "tutorial mode."

As you complete practice test 1, you should pay special attention to the tips located in the special boxes.

Although you will not see tips like this on the actual test, these suggestions will help you improve your exam performance.

You should also study the explanations to the answers to practice test 1 especially carefully.

The tips that you will see in the questions and explanations to math practice test 1 will help you obtain strategies to improve your performance on the other practice tests in this book.

Of course, these strategies will also help you do your best on the day of your actual CBEST math test.

CBEST Practice Math Test 1

1. Use the table to answer the question that follows.

Skill Area	Total Possible Points	Points Received
Flexibility	15	7
Strength	30	28
Speed	35	31
Stamina	20	16

An athlete takes part in a gymnastics competition and receives the points indicated in the table above. What percent of the total possible points did the athlete receive?

A. 28

B. 31

C. 72

D. 82

E. 92

> Question 1 is a numerical and graphic relationship question which involves using data presented in a table format. You have to decide which data to use from the table in order to solve the problem.

Tips and Explanations:

1. The correct answer is D.

STEP 1: First of all, you should add up the total possible points as shown.

15 + 30 + 35 + 20 = 100 possible points

STEP 2: Then add up the points the athlete received.

7 + 28 + 31 + 16 = 82 points received

STEP 3: Finally, divide the points received by the possible points to get the percentage of the total.

82 ÷ 100 = 82%

2. Two people are going to give money to a foundation for a project. Person A will provide one-half of the money. Person B will donate one-eighth of the money. What fraction represents the unfunded portion of the project?

   A. $^{1}/_{16}$
   B. $^{1}/_{8}$
   C. $^{1}/_{4}$
   D. $^{5}/_{8}$
   E. $^{3}/_{8}$

3. What is the lowest common denominator for the following equation?

   $$\left(\frac{1}{3}+\frac{11}{5}\right)+\left(\frac{1}{15}-\frac{4}{5}\right)$$

   A. 3
   B. 5
   C. 15
   D. 45
   E. 75

Questions 2 and 3 are computation and problem solving questions on performing calculations on fractions.

Tips and Explanations:

2. The correct answer is E.

You will see practical problems involving fractions like this one on the exam.

The sum of all contributions must be equal to 100%, simplified to 1.

STEP 1: Set up an equation. Let's say that the variable *U* represents the unfunded portion of the project.

So, the equation that represents this problem is $A + B + U = 1$

STEP 2: Substitute with the fractions that have been provided.

$$\frac{1}{2} + \frac{1}{8} + U = 1$$

STEP 3: Find the lowest common denominator (LCD). Finding the lowest common denominator means that you have to make all of the numbers on the bottoms of the fractions the same. Remember that you need to find the common factors of the denominators in order to find the LCD.

We know that 2 and 4 are factors of 8 because $2 \times 4 = 8$.

So, the LCD for this question is 8 since the denominator of the first fraction is 2 and because 2 is a factor of 8.

STEP 4: Convert the fractions into the lowest common denominator to solve the problem. We put the fractions into the LCD as follows:

$$\frac{1}{2}+\frac{1}{8}+U=1$$

$$\left(\frac{1}{2}\times\frac{4}{4}\right)+\frac{1}{8}+U=1$$

$$\frac{4}{8}+\frac{1}{8}+U=1$$

$$\frac{5}{8}+U=1$$

$$\frac{5}{8}-\frac{5}{8}+U=1-\frac{5}{8}$$

$$U=1-\frac{5}{8}$$

$$U=\frac{8}{8}-\frac{5}{8}$$

$$U=\frac{3}{8}$$

3. The correct answer is C.

We have to find the lowest common denominator (LCD) of the fractions. The LCD for this question is 15. We know this because the product of the other denominators is 3 times 5, which is 15.

We can illustrate the solution as follows:

$$\left(\frac{1}{3}+\frac{11}{5}\right)+\left(\frac{1}{15}-\frac{4}{5}\right)=$$

$$\left[\left(\frac{1}{3}\times\frac{5}{5}\right)+\left(\frac{11}{5}\times\frac{3}{3}\right)\right]+\left[\frac{1}{15}-\left(\frac{4}{5}\times\frac{3}{3}\right)\right]=$$

$$\frac{5}{15}+\frac{33}{15}+\frac{1}{15}-\frac{12}{15}$$

4. A hockey team had 50 games this season and lost 20 percent of them. How many games did the team win?

A. 8

B. 10

C. 20

D. 18

E. 40

Question 4 is a computation and problem solving question on performing calculations on percentages.

Tips and Explanations:

4. The correct answer is E.

For practical problems like this, you must first determine the percentage and formula that you need in order to solve the problem. Then, you must do long multiplication to determine how many games the team won.

Be careful. The question tells you the percentage of games the team lost, not won.

STEP 1: First of all, we have to calculate the percentage of games won.

If the team lost 20 percent of the games, we know that the team won the remaining 80 percent.

STEP 2: Now do the long multiplication.

$$\begin{array}{rl} 50 & \text{games in total} \\ \underline{\times\ .80} & \text{percentage of games won (in decimal form)} \\ 40.0 & \text{total games won} \end{array}$$

5. Carmen wanted to find the average of the five tests she has taken this semester. However, she erroneously divided the total points from the five tests by 4, which gave her a result of 90. What is the correct average of her five tests?

A. 64

B. 72

C. 80

D. 90

E. 110

> Question 5 is a question on statistical principles that involves performing arithmetic on basic data related to test scores and determining averages.

Tips and Explanations:

5. The correct answer is B.

STEP 1: First you need to find the total points that the student earned. You do this by taking Carmen's erroneous average times 4.

$4 \times 90 = 360$

STEP 2: Then you need to divide the total points earned by the correct number of tests in order to get the correct average.

$360 \div 5 = 72$

6. Estimate the result of the following: $502 \div 49.1$

A. 8

B. 9

C. 10

D. 11

E. 12

Question 6 is an estimation and measurement question that requires you to estimate the result of a problem involving division without doing the full computation.

Tips and Explanations:

6. The correct answer is C.

STEP 1: When doing estimation problems, you need to round the numbers up or down.

As a rule of thumb, numbers less than 5 will be rounded down to the nearest 0 and numbers of 5 or more will be rounded up to the nearest 10.

Our problem was $502 \div 49.1$

So, 502 is rounded down to 500 and 49.1 is rounded up to 50.

STEP 2: To estimate the result, we then perform the operation on the rounded figures.

$500 \div 50 = 10$

7. Which of the following is the greatest?

A. 0.540

B. 0.054

C. 0.045

D. 0.5045

E. 0.0054

Question 7 is a numerical and graphic relationship question that requires you to determine if a given number is less than or greater than other numbers. In this problem, the numbers are provided in decimal format.

Tips and Explanations:

7. The correct answer is A.

Put in extra zeroes and line up the decimal points in a column in order to compare the numbers like this:

0.5400

0.0540

0.0450

0.5045

If you are still not sure of your answer, you can remove the decimals as shown below to help you see the answer more clearly.

5400

540

450

5045

Therefore, the largest number is .540

8. Which of the following is the most appropriate unit of measurement for the weight of a car?

A. liters

B. horsepower

C. gallons

D. pounds per square inch

E. tons

Question 8 is an estimation and measurement question on understanding how to measure solid and fluid weights.

Tips and Explanations:

8. The correct answer is E.

You will need to understand the basic concepts of United States' measurements for the exam.

Remember that wet items are measured in pints and quarts, while dry items are measured in ounces and pounds, or tons in the case of extremely heavy quantities.

Feet and inches are linear measurements; they are not used for weight.

Liters and gallons are measures of liquid substances. Horsepower measures the strength of an engine.

Tons measure the weight of heavy items, so it would be suitable for measuring the weight of a car.

Note that one ton is equal to two thousand pounds.

9. Farmer Brown has a field in which cows craze. He is going to buy fence panels to put up a fence along one side of the field. Each panel is 8 feet 6 inches long. He needs 11 panels to cover the entire side of the field. How long is the field?

A. 60 feet 6 inches

B. 72 feet 8 inches

C. 93 feet 6 inches

D. 102 feet 8 inches

E. 110 feet 6 inches

Question 9 is an estimation and measurement question on measuring length and perimeter.

Tips and Explanations:

9. The correct answer is C.

Each panel is 8 feet 6 inches long, and he needs 11 panels to cover the entire side of the field. So, we need to multiply 8 feet 6 inches by 11, and then simplify the result.

Step 1: 8 feet × 11 = 88 feet

Step 2: 6 inches × 11 = 66 inches

There are 12 inches in a foot, so we need to determine how many feet and inches there are in 66 inches.

66 inches ÷ 12 = 5 feet 6 inches

Step 3: Now add the two results together.

88 feet + 5 feet 6 inches = 93 feet 6 inches

10. Marta uses one jar of coffee every 6 days. Approximately how many jars of coffee does she use per month?

A. 2

B. 3

C. 5

D. 6

E. 7

Question 10 is a problem solving question on performing calculations on whole numbers.

Tips and Explanations:

10. The correct answer is C.

There are 30 or 31 days in most months, so we need to take the number of days that it takes Marta to use one jar of coffee (which is 6 days in this problem) and divide that number into 30.

30 ÷ 6 = 5 jars per month

11. Jonathan can run 3 miles in 25 minutes. If he maintains this pace, how long will it take him to run 12 miles?

A. 1 hour and 15 minutes

B. 1 hour and 40 minutes

C. 1 hour and 45 minutes

D. 3 hours

E. 5 hours

> Question 11 is another problem solving question. This time, you have to solve a practical problem that involves both division and multiplication.

Tips and Explanations:

11. The correct answer is B.

STEP 1: Look to see what information is common to both the question and to the information provided. Here we have the information that he can run 3 miles in 25 minutes. The question is asking how long it will take him to run 12 miles, so the commonality is miles.

STEP 2: Next, you need to find out how many 3-mile increments there are in 12 miles.

$12 \div 3 = 4$

STEP 3: Then you need to determine the time required to travel the stated distance.

Accordingly, we need to multiply the time for 3 miles by 4.

25 minutes × 4 = 100

So, 100 minutes are needed to run 12 miles.

STEP 4: Finally, simplify into hours and minutes based on the fact that there are 60 minutes in one hour.

100 minutes = 1 hour 40 minutes.

12. A census shows that 1,008,942 people live in New Town, and 709,002 people live in Old Town. Which of the following numbers is the best estimate of how many more people live in New Town than in Old Town?

A. 330,000

B. 300,000

C. 33,000

D. 30,000

E. 3,000

Question 12 is another estimation and measurement problem involving estimating the results of problems without doing the full computation. To solve question 12, we need to perform subtraction.

Tips and Explanations:

12. The correct answer is B.

    As stated above, this is another type of estimation question.

    The problem tells us that 1,008,942 people live in New Town, and 709,002 people live in Old Town.

    STEP 1: We need to round the numbers up or down to the nearest thousand as needed.

    1,008,942 is rounded to 1,009,000

    709,002 is rounded to 709,000

    STEP 2: Then subtract the second figure from the first figure in order to get your result.

    1,009,000 – 709,000 = 300,000

13. Anne has taken a standardized college entrance exam. Use the report of her test scores below to answer the question that follows.

Raw Score Part 1	Raw Score Part 2	Mean	Standard Deviation	Percentile
180	230	205	15	78

Which of the following is a correct interpretation of the score report given above?

A. Ann scored as well as or better than 78% of the test takers.

B. Ann scored as well as or better than 85% of the test takers.

C. 15% of the test takers scored higher than Ann.

D. Ann answered 205 of the questions correctly.

E. Ann will perform well at college.

> Question 13 is a statistical principles problem on interpreting standardized test scores and understanding how a particular student has performed relative to other students.

Tips and Explanations:

13. The correct answer is A.

The raw score represents the number of questions that were answered correctly.

The mean is the average of the first two raw scores. In other words, we can calculate the mean like this: (180 + 230) ÷ 2 = 205

Standard deviation measures the variation from the mean or average.

The percentile rank of a score is the percentage of test-takers that scored the same or lower than the student in question. For instance, a percentile score of 60 means that 60% of the test-takers scored the same or lower than a particular student.

In our question, Anne's scores were in the 78th percentile, so Ann scored as well as or better than 78% of the test takers.

14. Simplify the following mathematical expression: –183 + 56 + (–17)

    A. 110

    B. 144

    C. –110

    D. –144

    E. –256

Question 14 is a computation problem involving both positive and negative numbers.

Tips and Explanations:

14. The correct answer is D.

    Be careful with negative signs when answering questions like this one. You might want to add the negatives together before adding in the positive numbers.

    –183 + 56 + (–17) = ?

    STEP 1: Deal with the negative numbers.

    –183 – 17 = –200

    STEP 2: Add in the positive number by reducing it from the negative number.

    –200 + 56 = –144

15. In a high school, 17 out of every 20 students participate in a sport. If there are 800 students at the high school, what is the total number of students that participate in a sport?

A. 120 students

B. 640 students

C. 680 students

D. 776 students

E. 780 students

Question 15 is another problem solving question. You will have to identify the required facts in the problem and then perform both division and multiplication in order to find the solution.

Tips and Explanations:

15. The correct answer is C.

Remember that for questions like this one, you have to find the commonality between the facts in the question and the requested information for the solution.

In this question, the commonality is the number of students.

The question tells us that 17 out of every 20 students participate in a sport and that there are 800 total students.

STEP 1: Determine how many groups of 20 can be formed from the total of 800.

$800 \div 20 = 40$ groups of 20 students in the school

STEP 2: To solve the problem, you then need to multiply the number of participants per group by the possible number of groups.

In this problem, there are 17 participants per every group of 20.

There are 40 groups of 20.

So, we multiply 17 by 40 to get our answer.

$17 \times 40 = 680$ students

16. The owner of a carnival attraction launches toy boats of different colors. There are 15 boats in total: 5 are blue, 3 are red, and 7 are green. If the first boat launched is green, and the next boat launched is selected at random, what is the probability that the next boat is blue? Note that a boat cannot be launched more than one time.

A. $^5/_{14}$

B. $^6/_{14}$

C. $^5/_{15}$

D. $^4/_{15}$

E. $^{10}/_{15}$

Question 16 is a statistical principles problem that involves understanding basic probability in order to make predictions based on the data provided.

Tips and Explanations:

16. The correct answer is A.

For questions on probability like this one, you need to reduce the quantity of the total data set by the quantity of items used.

STEP 1: Determine the total amount in the data set before any items are removed.

There are 15 boats in total before the first boat is launched.

5 + 3 + 7 = 15

STEP 2: Determine the numbers of items in the data set after items have been removed.

One boat is launched, so the amount in the data set is now 14.

In other words, the new data set is 14 since 15 – 1 = 14

STEP 3: Determine the amount in the subset.

The first boat launched is green, so we need to reduce the subset for that particular color by 1.

The new total for the green subset becomes 6 since 7 – 1 = 6

However, the question is asking us about the probability that the second boat will be blue, not green.

There are 5 blue boats, and a blue boat has not been launched so far.

STEP 4: The probability is expressed as a fraction.

The amount in the subset (5 blue boats) goes on the top of the fraction and the amount of items left in the data set (14 boats left) goes on the bottom.

So, the answer is $^5/_{14}$.

17. A new skyscraper is being erected in the city center. The foundation of the building extends 1,135 feet below ground. The building itself, when erected, will measure 13,975 feet above ground. Which of the following is the best estimate of the distance between the deepest point of the foundation below ground and the top of the erected building above ground?
    A. 12,000 feet
    B. 13,000 feet
    C. 14,000 feet
    D. 15,000 feet
    E. 16,000 feet

> Question 17 is an estimation problem that involves a measurement below ground and another measurement above ground.

Tips and Explanations:

17. The correct answer is D.

    The foundation of the building extends 1,135 feet below ground. The building itself, when erected, will measure 13,975 feet above ground.

STEP 1: Look at the answer options in order to determine what increments are required.

Here, we see that the answer options are in increments of one thousand.

STEP 2: Perform the rounding on both figures.

So, we round each number up or down to the nearest thousand.

1,135 is rounded down to 1,000

13,975 is rounded up to 14,000

STEP 3: Perform the necessary mathematical computation.

We add the two figures together from above in order to get our result.

1,000 + 14,000 = 15,000

18. Mrs. Johnson is going to give candy to the students in her class. The first bag of candy that she has contains 43 pieces. The second contains 28 pieces, and the third contains 31 pieces. If there are 34 students in Mrs. Johnson's class, and the candy is divided equally among all of the students, how many pieces of candy will each student receive?

A. 3 pieces
B. 4 pieces
C. 5 pieces
D. 51 pieces
E. 102 pieces

Question 18 is a computation and problem solving question that involves both addition and division.

Tips and Explanations:

18. The correct answer is A.

STEP 1: First of all, we need to find out how many pieces of candy there are in total.

43 + 28 + 31 = 102 total pieces of candy

STEP 2: We need to divide the total amount of candy by the number of students in order to find out how much candy each student will get.

102 total pieces of candy ÷ 34 students = 3 pieces of candy per student

19. One hundred students took an English test. The 55 female students in the class had an average score of 87, while the 45 male students in the class had an average of 80. What is the average test score for all 100 students in the class?

A. 82.00

B. 83.15

C. 83.50

D. 83.85

E. 84.00

Question 19 is another statistical principles problem on performing operations on data relating to test scores and calculating averages.

Tips and Explanations:

19. The correct answer is D.

STEP 1: First of all, you have to calculate the total amount of points earned by the entire class.

Multiply the female average by the amount of female students.

Total points for female students: $87 \times 55 = 4785$

Then multiply the male average by the amount of male students.

Total points for male students: $80 \times 45 = 3600$

STEP 2: Then add these two amounts together to find out the total points scored by the entire class.

Total points for entire class: $4785 + 3600 = 8385$

STEP 3: When you have calculated the total amount of points for the entire class, you divide this by the total number of students in the class to get the class average. There are 100 total students in this class because there are 55 females and 45 males.

$8385 \div 100 = 83.85$

20. Two people are going to work on a job. The first person will be paid \$7.25 per hour. The second person will be paid \$10.50 per hour. A represents the number of hours the first person will work, and B represents the number of hours the second person will work.

What equation represents the total cost of the wages for this job?

A. 17.75AB

B. 17.75 ÷ AB

C. AB ÷ 17.75

D. (7.25A + 10.50B)

E. (10.50A + 7.25B)

Question 20 is a computation and problem solving question involving the use of an algebraic equation in order to solve a practical problem.

Tips and Explanations:

20. The correct answer is D.

STEP 1: Assign variables as necessary. The two people are working at different costs per hour, so each person needs to be assigned a variable. A is for the number of hours for the first person, and B is for the number of hours for the second person.

STEP 2: The cost for each person is calculated by taking the number of hours that the person works by the hourly wage for that person.

So, the equation for wages for the first person is (7.25 × A)

The equation for the wages for the second person is (10.50 × B)

STEP 3: The total cost of the wages for this job is the sum of the wages of these two people.

$(7.25 \times A) + (10.50 \times B) =$

$(7.25A + 10.50B)$

21. If $5x - 4(x + 2) = -2$, then $x = ?$

A. 0

B. 8

C. 6

D. –8

E. –6

Question 21 is a computation problem on solving algebraic equations with one unknown variable.

Tips and Explanations:

21. The correct answer is C.

In order to solve algebraic equations with one unknown variable, you have to multiply and then isolate the *x* variable

STEP 1: Perform the multiplication on the parenthetical expression.

$5x - 4(x + 2) = -2$

$5x - 4x - 8 = -2$

STEP 2: Then perform any other operations, such as addition or subtraction.

$(5x - 4x) - 8 = -2$

$x - 8 = -2$

STEP 3: Deal with the remaining whole number.

$x - 8 = -2$

$x - 8 + 8 = -2 + 8$

STEP 4: Isolate the variable to solve the problem

$x - 8 + 8 = -2 + 8$

$x - 0 = -2 + 8$

$x = -2 + 8$

$x = 6$

22. The Johnsons have decided to remodel their upstairs. They currently have 4 rooms upstairs that measure 10 feet by 10 feet each. When they remodel, they will make one large room that will be 20 feet by 10 feet and two small rooms that will each be 10 feet by 8 feet. The remaining space is to be allocated to a new bathroom. What are the dimensions of the new bathroom?

A. 4 feet × 10 feet

B. 8 feet × 10 feet

C. 10 feet × 10 feet

D. 4 feet × 8 feet

E. 8 feet × 8 feet

Question 22 is a more advanced measurement problem that involves length, width, and total linear dimensions.

Tips and Explanations:

22. The correct answer is A.

STEP 1: First, we have to calculate the total square footage available. If there are 4 rooms which are 10 by 10 each, we have this equation:

$4 \times (10 \times 10) = 400$ square feet in total

STEP 2: Now calculate the square footage of the new rooms.

$20 \times 10 = 200$

2 rooms $\times (10 \times 8) = 160$

$200 + 160 = 360$ total square feet for the new rooms

STEP 3: The remaining square footage for the bathroom is calculated by taking the total minus the square footage of the new rooms.

$400 - 360 = 40$ square feet

Since each existing room is 10 feet long, we know that the new bathroom also needs to be 10 feet long in order to fit in. So, the new bathroom is 4 feet × 10 feet.

23. Use the table below to answer the following question:

Sunday	Monday	Tuesday	Wednesday	Thursday	Friday	Saturday
–10°F	–9°F	1°F	6°F	8°F	13°F	12°F

The weather forecast for the coming week is given in the table above. What is the difference between the highest and lowest forecasted temperatures for the week?

A. –2°F

B. –3°F

C. 3°F

D. 22°F

E. 23°F

> Question 23 is a numerical relationship problem that involves identifying and using information from a chart.

Tips and Explanations:

23. The correct answer is E.

The lowest temperature is –10°F, and the highest temperature is 13°F.

The difference between these two figures is calculated by subtracting.

Be careful when you subtract. In particular, remember that when you see two negative signs together, you need to add. In other words, two negatives make a positive.

$13 - (-10) =$

$13 + 10 = 23$

24. Acme Packaging uses string to secure their packages prior to shipment. The string is tied around the entire length and entire width of the package, as shown in the following illustration:

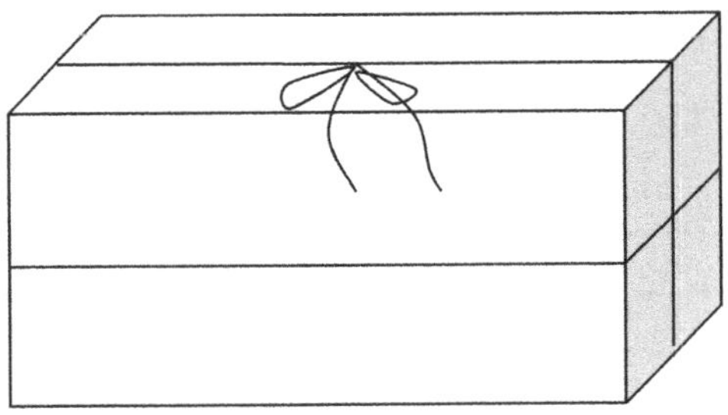

The box is ten inches in height, ten inches in depth, and twenty inches in length. An additional fifteen inches of string is needed to tie a bow on the top of the package. How much string is needed in total in order to tie up the entire package, including making the bow on the top?

A. 55 inches
B. 95 inches
C. 120 inches
D. 130 inches
E. 135 inches

Question 24 is an advanced measurement problem with a calculation involving length, width, and depth in inches.

Tips and Explanations:

24. The correct answer is E.

For questions that ask you about tying string around a package, you will need to consider the length, width, and depth of the package when doing your calculation.

STEP 1: The string that goes around the top, bottom, and ends of the package will be measured as follows: 20 + 10 + 20 + 10 = 60 inches

STEP 2: The string that goes around the front and back sides and the ends of the package will be calculated similarly since the front and back sides are of the course the same length as the top and bottom.

20 + 10 + 20 + 10 = 60 inches

STEP 3: Don't forget that an additional fifteen inches of rope is needed to tie a bow on the top of the package.

STEP 4: We add these three amounts together to get our total.

60 + 60 + 15 = 135 inches

25. Yesterday a train traveled $117^{3}/_{4}$ miles. Today it traveled $102^{1}/_{6}$ miles. What is the difference between the distance traveled today and yesterday?

A. 15 miles

B. $15^1/_4$ miles

C. $15^7/_{12}$ miles

D. $15^9/_{12}$ miles

E. $16^5/_6$ miles

Question 25 is a problem solving question that involves performing calculations on whole numbers and fractions.

Tips and Explanations:

25. The correct answer is C.

Yesterday the train traveled $117^3/_4$ miles, and today it traveled $102^1/_6$ miles. To find the difference, we subtract these two amounts.

Because the fraction on the first mixed number is greater than the fraction on the second mixed number, we can subtract the whole numbers and the fractions separately.

$117^3/_4$ miles – $102^1/_6$ miles = ?

STEP 1: Subtract the whole numbers.

117 – 102 = 15 miles

STEP 2: Perform the operation on the fractions by finding the lowest common denominator.

$^3/_4$ miles – $^1/_6$ miles = ?

In order to find the LCD, we would normally need to find the common factors first.

Our denominators in this problem are 4 and 6.

The factors of 4 are:

$1 \times 4 = 4$

$2 \times 2 = 4$

The factors of 6 are:

$1 \times 6 = 6$

$2 \times 3 = 6$

We do not have two factors in common, so we know that we need to find a new denominator which is greater than 6.

In this problem, the LCD is 12 since $3 \times 4 = 12$ and $2 \times 6 = 12$.

So, we express the fractions $^3/_4$ miles + $^1/_6$ miles from above in their LCD form.

$^3/_4 \times {^3/_3} = {^9/_{12}}$

$^1/_6 \times {^2/_2} = {^2/_{12}}$

Then subtract these two fractions.

$^9/_{12} - {^2/_{12}} = {^7/_{12}}$

STEP 3: Combine the results from the two previous steps to solve the problem.

$117^3/_4$ miles – $102^1/_6$ miles = 15 $^7/_{12}$ miles

26. Liz wants to put new vinyl flooring in her kitchen. She will buy the flooring in square pieces that measure 1 square foot each. The entire room is 8 feet by 12 feet. The cupboards are two feet deep from front to back. Flooring will not be put under the cupboards.

A diagram of her kitchen is provided.

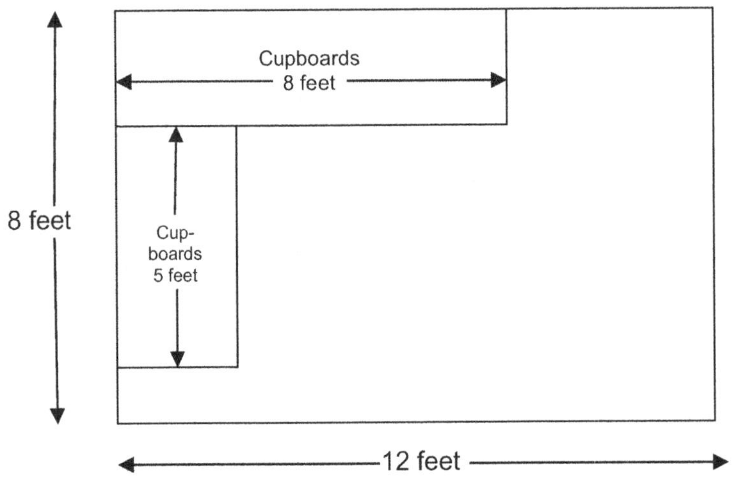

How many pieces of vinyl will Liz need to cover her floor?

A. 84

B. 70

C. 88

D. 96

E.120

> Question 26 is another advanced measurement problem. Be sure to read the facts provided in problems like this one very carefully.

Tips and Explanations:

26. The correct answer is B.

For problems like this one, find your solution for each part of the floor and then add these parts together.

STEP 1: First we will find the square footage of the shaded area below.

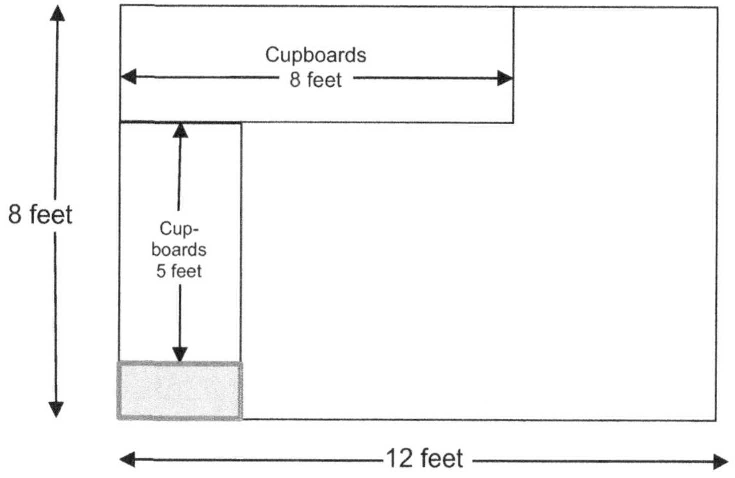

This problem is tricky because flooring is not being placed under the cupboards, so we have to find the square footage for an irregular area.

The cupboards are two feet deep, and the room is 8 feet along the side, so there is a remaining area along the cupboard here of 1 foot (8 feet minus 2 feet for the back cupboard minus 5 feet for the side cupboard = 1 foot) by 2 feet (since the cupboards are two feet deep).

1 foot × 2 feet = 2 square feet

STEP 2: Now find the square footage along the side of the other cupboard.

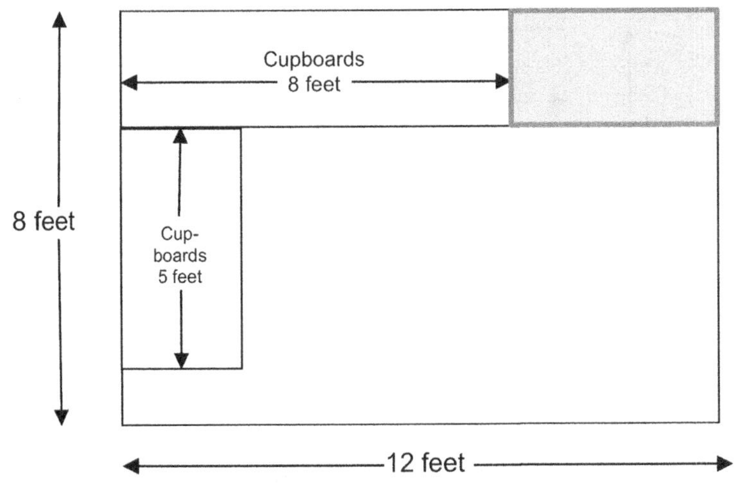

The room is 12 feet along the front, so there is a remaining area along the cupboard here of 4 feet (12 feet minus 8 feet for the length of the back cupboard = 4 feet) by 2 feet (since the cupboards are two feet deep). 4 feet × 2 feet = 8 square feet

STEP 3: Find the square footage for the remaining floor area.

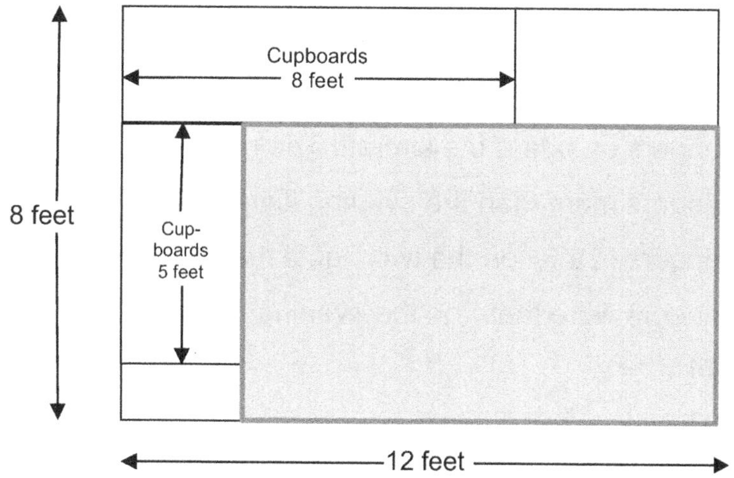

The room is 12 feet along the front and the cupboards are two feet deep, so there is a remaining floor area along the front of 10 feet (12 feet minus 2 feet for the depth of the side cupboard = 10 feet).

The room is 8 feet along the side and the cupboards are two feet deep, so there is a remaining floor area along the side here of 6 feet (8 feet minus 2 feet for the depth of the back cupboard = 10 feet).

So the remaining area here is 60 square feet.

10 feet × 6 feet = 60 square feet

STEP 4: Now add the three results together.

Result from step 1: 1 foot × 2 feet = 2 square feet

Result from step 2: 4 feet × 2 feet = 8 square feet

Result from step 3: 10 feet × 6 feet = 60 square feet

2 + 8 + 60 = 70 square feet

27. During each flight, a flight attendant must count the number of passengers on board the aircraft. The morning flight had 52 passengers more than the evening flight, and there were 540 passengers in total on the two flights that day. How many passengers were there on the evening flight?

A. 244

B. 296

C. 488

D. 540

E. 592

Question 27 is a computation and problem solving question that involves allocating more items to one part of the total than to the other part of the total.

Tips and Explanations:

27. The correct answer is A.

The problem tells us that the morning flight had 52 passengers more than the evening flight, and there were 540 passengers in total on the two flights that day.

STEP 1: First of all, we need to deduct the difference from the total:

540 – 52 = 488

In other words, there were 488 passengers on both flights combined, plus the 52 additional passengers on the morning flight.

STEP 2: Now divide this result by 2 to allocate an amount of passengers to each flight.

488 ÷ 2 = 244 passengers on the evening flight

So, the evening flight had 244 passengers

Had the question asked you for the amount of passengers on the morning flight, you would have had to add back the amount of additional passengers to find the total amount of passengers for the morning flight.

244 + 52 = 296 passengers on the morning flight

28. Sam is driving a truck at 70 miles per hour. At 10:30 am, he sees this sign:

Brownsville	**35 miles**
Dunnstun	**70 miles**
Farnam	**140 miles**
Georgetown	**210 miles**

After Sam sees the sign, he continues to drive at the same speed. At 11:00 am, how far will he be from Farnam?

A. He will be in Farnam.

B. He will be 35 miles from Farnam.

C. He will be 70 miles from Farnam.

D. He will be 105 miles from Farnam.

E. He will be 175 miles from Farnam.

Question 28 is a practical problem solving question on calculating distance traveled.

Tips and Explanations:

28. The correct answer is D.

Sam is driving at 70 miles per hour, and at 10:30 am he is 140 miles from Farnam.

STEP 1: We need to find out how far he will be from Farnam at 11:00 am, so we need to work out how far he will travel in 30 minutes.

STEP 2: If Sam is traveling at 70 miles an hour, then he travels 35 minutes in half an hour.

70 miles in one hour × $^1/_2$ hour = 35 miles

STEP 3: If he was 140 miles from Farnam at 10:30 am, he will be 105 miles from Farnam at 11:00 am.

140 – 35 = 105 miles

29. In a math class, $^1/_3$ of the students fail a test. If twelve students have failed the test, how many students are in the class in total?

A. 15

B. 16

C. 36

D. 38

E. 48

> Question 29 is an advanced practical problem solving question involving the use of fractions.

Tips and Explanations:

29. The correct answer is C.

The twelve students who failed the test represent one-third of the class. Since one-third of the students have failed, we can think of the class as being divided into three groups:

Group 1: The 12 students who failed

Group 2: 12 students who would have passed

Group 3: 12 more students who would have passed

So, the class consists of 36 students in total.

In other words, we need to multiply by three to find the total number of students.

$12 \times 3 = 36$

30. Mark owns a bargain bookstore that sells every book for $5. Last week, his sales were $525. This week his sales figure was $600.

How many more books did Mark sell this week, compared to last week?

A. 5

B. 15

C. 25

D. 75

E. 105

Question 30 is a practical problem involving the calculation of the number of units sold.

Tips and Explanations:

30. The correct answer is B.

The problem tells us that sales this week were \$600 and sales last week were \$525.

STEP 1: First, we need to find the difference in sales between the two weeks.

\$600 - \$525 = \$75 more in sales this week

STEP 2: Since each book is sold for \$5, we divide this figure into the total in order to find out how many books were sold.

\$75 more sales ÷ \$5 per book = 15 more books sold this week

31. A candy store sells chocolate candy bars. At the beginning of the day, it has 60 candy bars available for sale in total. 25 of them are milk chocolate, 20 are white chocolate, and 15 are dark chocolate. At

the close of business that day, 7 candy bars have not been sold. Which of the following details can be determined from the information given above?

A. average sales of candy bars that week
B. amount of white dark chocolate bars sold that day
C. total amount of candy bars sold that day
D. which candy bar is the most popular normally
E. the difference between the amount of milk chocolate and dark chocolate candy bars sold that day

> Question 31 is a computation and problem solving question on determining if enough information is provided in order to solve a problem.

Tips and Explanations:

31. The correct answer is C.

For problems like this, you will need to read the information provided in the problem carefully and then rule out the incorrect answers one by one.

The problem does not tell us the amount of candy bars sold on other days of the week or year, so we cannot calculate an average for the week as stipulated in answer choice A, nor can we determine which candy bar is the most popular normally, as stipulated in answer choice D.

The problem also does not tell us the amount of each type of chocolate bar sold. We do not know how many of the candy bars sold were milk chocolate, nor do we know how many were white or dark chocolate. Therefore, we cannot determine the information requested in answer choices B and E.

We can determine the amount of candy bars sold that day as stated in answer choice C since we know that there were 60 candy bars at the beginning of the day and 7 have not been sold, so 53 candy bars were sold that day.

32. The price of socks is $2 per pair and the price of shoes is $25 per pair. Anna went shopping for socks and shoes, and she paid $85 in total. In this purchase, she bought 3 pairs of shoes. How many pairs of socks did she buy?

A. 2

B. 3

C. 5

D. 8

E. 15

Question 32 is a practical problem solving question involving calculations on prices per unit.

Tips and Explanations:

32. The correct answer is C.

Assign a different variable to each item, and then make an equation by multiplying each variable by its price.

STEP 1: Assign the variables.

Let's say that the number of pairs of socks is S and the number of pairs of shoes is H.

STEP 2: Set up your equation.

Your equation is: $(S \times \$2) + (H \times \$25) = \$85$

STEP 3: We know that the number of pairs of shoes is 3, so put that in the equation and solve it.

$(S \times \$2) + (H \times \$25) = \$85$

$(S \times \$2) + (3 \times \$25) = \$85$

$(S \times \$2) + \$75 = \$85$

$(S \times \$2) + 75 - 75 = \$85 - \$75$

$(S \times \$2) = \$10$

$\$2S = \$10$

$\$2S \div 2 = \$10 \div 2$

$S = 5$

33. Which of the following mathematical expressions equals $^3/_{xy}$?

A. $^3/_x \times ^3/_y$

B. $3 \div 3xy$

C. $3 \div (xy)$

D. $^1/_3 \div 3xy$

E. $^1/_3 \div (x3y)$

Question 33 is a numerical relationship problem on identifying mathematical equivalents.

Tips and Explanations:

33. The correct answer is C.

When you see a fraction, the line in the fraction can be treated as the division symbol. For example, $^3/_5 = 3 \div 5$

Using the same principle, $^3/_{xy} = 3 \div (xy)$

34. Which of the following numbers is between 4,789,321 and 4,901,312?

A. 4,587,624

B. 4,780,201

C. 4,789,231

D. 4,789,320

E. 4,792,558

Questions 34 and 35 are numerical relationship problems on determining if a given number is less than or greater than other numbers.

Tips and Explanations:

34. The correct answer is E.

For questions like this one, you can line up all of the seven figures in a column in order to compare them.

4,789,321

A. 4,587,624

B. 4,780,201

C. 4,789,231

D. 4,789,320

E. 4,792,558

4,901,312

Comparing each digit of each figure as we go along, we can see that answer choices A to D are each less than 4,789,321. Answer choice E is greater than 4,789,321 and less than 4,901,312, so it is correct.

35. If the value of x is between 0.0007 and 0.0021, which of the following could be x?

A. 0.0012

B. 0.0006

C. 0.0022

D. 0.022

E. 0.08

35. The correct answer is A.

This problem is like the previous one, except this question involves decimals. For problems with decimals, line the figures up in a column and add zeroes to fill in the column as shown.

0.0007

A. 0.0012

B. 0.0006

C. 0.0022

D. 0.0220

E. 0.0800

0.0021

Answer choice B is less than 0.0007, and answer choices C, D, and E are greater than 0.0021.

Answer choice A (0.0012) is between 0.0007 and 0.0021, so it is the correct answer.

36. Terry runs 9 miles every day. If his daily run is rounded up to the nearest 5 miles, which of the following is the best estimate of how many miles he runs every 5 days?

A. 25

B. 35

C. 45

D. 50

E. 70

Question 36 is an estimation problem involving the use of rounding in order to find the distance traveled.

Tips and Explanations:

36. The correct answer is D.

STEP 1: Looking at the answer choices, we can see that we need to round to the nearest increment of 5.

So, for this problem, think about the increments of 5:

5, 10, 15, 20, 25, etc.

STEP 2. Perform the rounding.

9 miles per day is rounded up to 10 miles per day.

STEP 3: Multiply to find the solution.

We then multiply this figure by the number of days to get our result.

10 miles per day × 5 days = 50 miles every five days

37. Kathy is on a diet. During week 1, she lost 1.07 pounds. During week 2, she lost 2.46 pounds. During week 3, she lost 3.92 pounds. If each week's weight loss amount is rounded up or down to the nearest one-tenth of a pound, what is the estimate of Kathy's weight loss for the entire 3 weeks?

A. 7 pounds
B. 7.40 pounds
C. 7.45 pounds
D. 7.50 pounds
E. 8 pounds

Question 37 is an estimation problem involving the rounding of decimals.

Tips and Explanations:

37. The correct answer is D.

One-tenth is expressed is decimal form like this: 0.1

So, any amount that has a decimal less than 0.05 is rounded down and decimals of 0.05 and greater are rounded up.

STEP 1: We do the rounding as follows:

1.07 is rounded up to 1.1

2.46 is rounded up to 2.5

3.92 is rounded down to 3.9

STEP 2: Then add the rounded figures together to get your result.

$1.1 + 2.5 + 3.9 = 7.5$

38. Use the information provided in the box below to answer the question that follows.

- The police station is 10 miles away from the fire station
- The fire station is 6 miles away from the hospital.

Based on the information in the box, what conclusions can be made?

A. The police station is no more than 6 miles away from the hospital.

B. The police station is no more than 10 miles away from the hospital.

C. The police station is exactly 6 miles away from the hospital.

D. The fire station is exactly 10 miles away from the hospital.

E. The police station is no more than 16 miles away from the hospital.

Question 38 is a numerical and graphic relationship word problem on logical relationships.

Tips and Explanations:

38. The correct answer is E.

For questions about distance like this one, keep in mind that the locations may or may not lie on a straight line.

For example, the locations could be laid out like this:

In the layout above, the police station would be 16 miles from the hospital.

However, the locations could also be laid out like this:

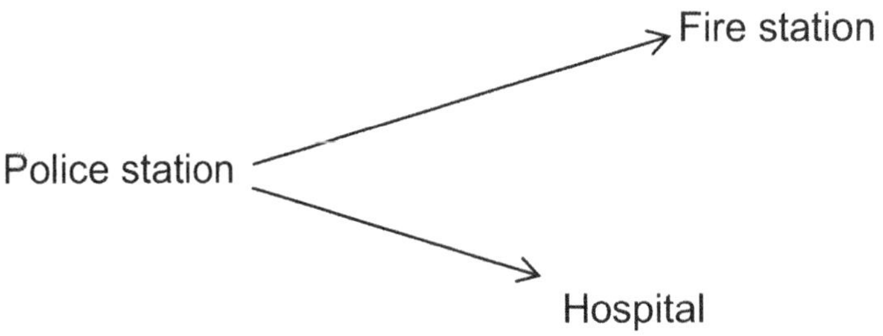

We can see that the locations will be the farthest from each other if they are laid out on a straight line as in the first example above.

In other words, a person could always go to the hospital by traveling to the fire station from the police station (10 miles) and then traveling from the fire station to the hospital (6 miles).

Therefore, the police station would never be more than 16 miles away from the hospital, regardless of the layout.

39. Kieko needs to calculate 16% of 825. Which of the following formulas can she use?

A. 825×16

B. 16×825

C. 825×16

D. 825×1.6

E. 825×0.16

Question 39 is a computation and problem solving question on identifying mathematical equivalents of percentages and decimals.

Tips and Explanations:

39. The correct answer is E.

A percentage can always be expressed as a number with two decimal places.

For example, 15% = 0.15 and 20% = 0.20

In our problem, 16% = 0.16

Therefore, E is the correct answer.

40. Wei Lei bought a shirt on sale. The original price of the shirt was $18, and he got a 40% discount. What was the sales price of the shirt?

A. $7.20

B. $10.80

C. $11.80

D. \$17.28

E. \$17.60

Question 40 is another practical problem on the calculation of discounts.

Tips and Explanations:

40. The correct answer is B.

STEP 1: First of all, you need to calculate the amount of the discount.

\$18 original price × 40% =

\$18 × 0.40 = \$7.20 discount

STEP 2: Then deduct the amount of the discount from the original price to calculate the sales price of the item.

\$18 original price - \$7.20 discount = \$10.80 sales price

41. Professor Smith uses a system of extra-credit points for his class. Extra-credit points can be offset against the points lost on an exam due to incorrect responses. David answered 18 questions incorrectly on the exam and lost 36 points. He then earned 25 extra credit points. By how much was his exam score ultimately lowered?

A. −11

B. 11

C. 18

D. 25

E. 36

Question 41 is another problem on performing arithmetic on data relating to test scores. It involves operations with both positive and negative numbers.

Tips and Explanations:

41. The correct answer is B.

If David answered 18 questions incorrectly on the exam and lost 36 points, and he then earned 25 extra credit points, his score was lowered by 11 points.

STEP 1: To do the calculation, we need to take the points lost on the exam and add the extra credit points.

$-36 + 25 = -11$

STEP 2: Since the question is asking how much the score was lowered, you need to give the amount as a positive number.

42. What number is next in this sequence? 2, 4, 8, 16

A. 18

B. 20

C. 24

D. 32

E. 36

Questions on sequences like question 42 are another type of numerical relationship problem.

Tips and Explanations:

42. The correct answer is D.

Try to find the pattern of relationship between the numbers.

Here, we can see this pattern:

$2 \times 2 = 4$

$4 \times 2 = 8$

$8 \times 2 = 16$

In other words, the next number in the sequence is always double the previous number.

Therefore the answer is: $16 \times 2 = 32$

43. The county is proposing a 7.5% increase in its annual real estate tax. If the tax is currently $480 per year, how much would the tax be if the proposed increase is approved?

A. $444

B. $487

C. $516

D. $840

E. $3330

> Question 43 is a practical problem solving question on using percentages to calculate a tax increase.

Tips and Explanations:

43. The correct answer is C.

STEP 1: Calculate the amount of the tax increase.

$\$480 \times 7.5\% = ?$

\$480 original tax amount × 0.075 = \$36 proposed increase in tax

STEP 2: Then add the increase to the original amount to get the amount of the tax after the proposed increase.

\$480 original tax + \$36 increase in tax = \$516 tax after increase

44. Which one of the values will correctly satisfy the following mathematical statement: $^{2}/_{3} < ? < ^{7}/_{9}$

A. $^{1}/_{3}$

B. $^{1}/_{5}$

C. $^{2}/_{6}$

D. $^{1}/_{2}$

E. $^{7}/_{10}$

Question 44 is a numerical relationship problem on ordering fractions from least to greatest.

Tips and Explanations:

44. The correct answer is E.

This is another question involving common denominators.

The question is: $^{2}/_{3} < ? < ^{7}/_{9}$

STEP 1: First of all, we need to find a common denominator for the fractions in the equations, as well as for all of the answer choices. In order to complete the problem quickly, you should not try to find the lowest common denominator, but just find any common denominator. We can do this by expressing all of the numbers with a denominator of 90 since 9 is the largest denominator in the equation and 10 is the largest denominator in the answer choices.

$^{2}/_{3} \times ^{30}/_{30} = ^{60}/_{90}$

$^{7}/_{9} \times ^{10}/_{10} = ^{70}/_{90}$

STEP 2: Then, express the original equation in terms of the common denominator.

$^{60}/_{90} < ? < ^{70}/_{90}$

STEP 3: Then express the answer choices in terms of the common denominator.

A. $^{1}/_{3} \times ^{30}/_{30} = ^{30}/_{90}$

B. $^{1}/_{5} \times ^{18}/_{18} = ^{18}/_{90}$

C. $^{2}/_{6} \times ^{15}/_{15} = ^{30}/_{90}$

D. $^{1}/_{2} \times ^{45}/_{45} = ^{45}/_{90}$

E. $^{7}/_{10} \times ^{9}/_{9} = ^{63}/_{90}$

STEP 4: Compare the results to find the answer.

By comparing the numerators (the top numbers of the fractions), we can see that $^{63}/_{90}$ lies between $^{60}/_{90}$ and $^{70}/_{90}$, so E is the correct answer.

45. Use the information in the box below to answer the question that follows.

An orchard grows apples for resale. If the apple is 8 inches or more around, it is classified as grade A and sold to exclusive retailers. If the apple measures less than 8 inches around, but more than 4 inches around, it is classified as grade B and sold to wholesalers. If the apple measures 4 inches or less around it is classified as grade C. Apples with a grade C classification are rejected for human consumption and are sold to animal food manufacturers.

If an apple measures exactly 4 inches around, which of the following statements could be true?

A. The apple will be classified as grade B.

B. The apple will be sold to exclusive retailers.

C. The apple will be sold to wholesalers.

D. The apple will not be classified as grade C.

E. The apple will not be eaten by people.

Questions 45 and 46 are further word problems on logical relationships

Tips and Explanations:

45. The correct answer is E.

Read the facts of problems like this one very carefully. The facts provided in the problem tell us that if an apple measures 4 inches or less around it is classified as grade C, which is sold to animal food manufacturers. The apple in this problem is exactly 4 inches, so it is a grade C apple. Therefore, the apple will not be eaten by people, but by animals.

46. Use the information in the box below to answer the question that follows.

School will be held every weekday from Monday to Friday from 15th August until 22nd December from 8:30 am to 3:00 pm. However, if the temperature is more than 100 degrees, school will be dismissed at 11:30 am. School will not be held on public holidays.

If Tom did not go to school today, then which of the following statements must be true?

A. It is the 22nd of August.

B. The temperature exceeds 100 degrees.

C. It is a public holiday or the temperature exceeds 100 degrees.

D. It is a public holiday and the temperature exceeds 100 degrees.
E. It is a public holiday or a Saturday or Sunday.

46. The correct answer is E.

This is another type of problem in which you have to assess the available facts, so read carefully and do not make any assumptions that are not supported by the information provided.

As stated previously, it is usually best to deal with each answer option one by one for these types of questions.

Answer choice A is incorrect. If it is the 22nd of August, Tom would be in school because school is held every weekday from Monday to Friday from 15th August until 22nd December.

Answer choices B, C, and D are incorrect because even if the temperature was in excess of 100 degrees, Tom would have attended school from 8:30 to 11:30 am.

Therefore, we know that answer E is the correct answer. We also know that E is correct because school is held only on weekdays, and answer E stipulates that it is a Saturday, Sunday, or public holiday.

47. Use the information in the chart below to answer the question that follows.

Fatal Traffic Accidents in Hawaii, Alaska, Texas, and the Other States				
Year	Hawaii	Alaska	Texas	Other States
1994	2,365	987	3,687	52,187
1999	1,987	882	3,522	48,233
2004	1,784	915	3,601	51,505
2009	1,801	899	3,547	50,689
2014	1,621	823	3,623	49,117

How many fatal traffic accidents occurred outside the state of Texas in 2004?

A. 2,699

B. 3,601

C. 51,505

D. 54,204

E. 55,539

Questions 47 to 50 are graphic relationship problems with tables, line graphs, and pie charts

Tips and Explanations:

47. The correct answer is D.

The question is asking you for the total of all states, besides Texas, so you need to add together the amounts for Hawaii, Alaska, and the remaining states for the year 2004.

1,784 + 915 + 51,505 = 54,204

48. Use the information in the graph below to answer the question that follows.

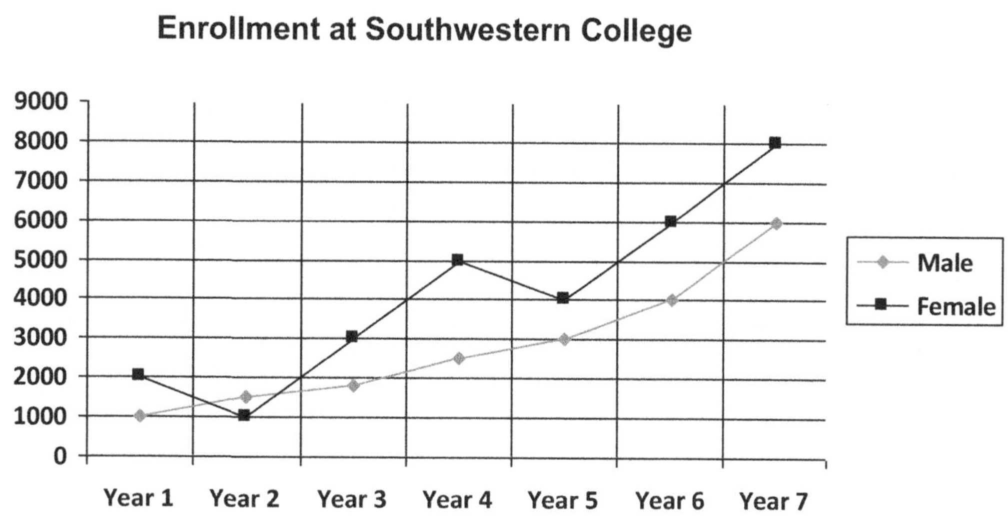

The chart above shows enrollment at Southwestern College over a seven-year period. According to the chart above, what was the largest approximate difference between male and female enrollment in any of the years displayed on the graph?

A. 300

B. 500

C. 1000

D. 2000

E. 2300

48. The correct answer is E.

For graphic questions like this one, you can usually find the answer by visually inspecting the graph. However, be sure to double-check your answer.

The points appear to be the farthest apart in year 4, where we see that the points are more than two lines apart.

The second farthest points are only two lines apart.

Each line represents one thousand students, so the difference between the amount of male and female students in year 4 is approximately 2,300 students.

49. Use the information in the graph below to answer the question that follows.

Favorite Drinks of Customers at Metro Bar and Grill

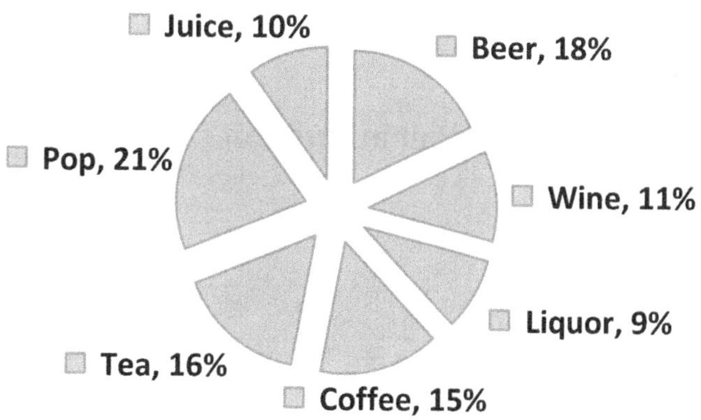

What percentage of customers at Metro Bar and Grill have a favorite drink that is a non-alcoholic beverage?

A. 62%

B. 48%

C. 41%

D. 42%

E. 23%

49. The correct answer is A.

Juice, pop, tea, and coffee are non-alcoholic beverages, so we need to add up these percentages.

10% + 21% + 16% + 15% = 62%

50. Use the information in the graph below to answer the question that follows.

Annual Rainfall in Hudson County

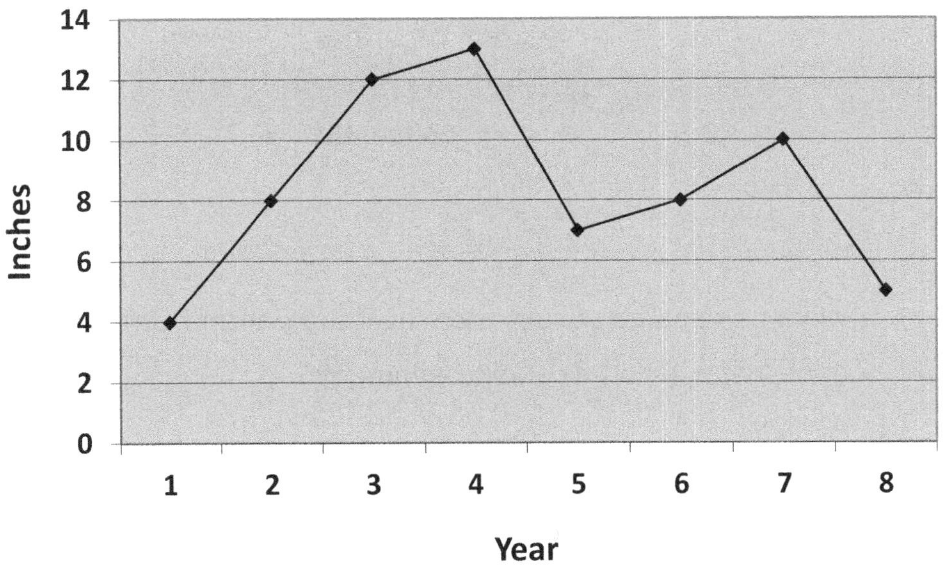

Between what two years did the amount of rainfall change by the greatest amount?

A. between year 1 and year 2

B. between year 2 and year 3

C. between year 4 and year 5

D. between year 5 and year 6

E. between year 7 and year 8

50. The correct answer is C.

The distance between each set of horizontal lines on the graph represents two inches of rain. We can see that the change is biggest between years 4 and 5, since the line of the graph stretches across three horizontal lines between these two years.

CBEST Practice Math Test 2

1. A group of friends are trying to lose weight. Person A lost $14^3/_4$ pounds. Person B lost $20^1/_5$ pounds. Person C lost 36.35 pounds. What is the total weight loss for the group?

   A. 70.475

   B. 71.05

   C. 71.15

   D. 71.25

   E. 71.30

2. The university bookstore is having a sale. Course books can be purchased for $40 each, or 5 books can be purchased for a total of $150. How much would a student save on each book if he or she purchased 5 books?

   A. 5

   B. 10

   C. 50

   D. 90

   E. 110

3. 120 students took a math test. The 60 female students in the class had an average score of 95, while the 60 male students in the class had an average of 90. What is the average test score for all 120 students in the class?

   A. 75

   B. 92.5

   C. 93

D. 93.5

E. 120

4. Tom bought a shirt on sale for $12. The original price of the shirt was $15. What was the percentage of the discount on the sale?

A. 2%

B. 3%

C. 20%

D. 25%

E. 30%

5. A car travels at 60 miles per hour. The car is currently 240 miles from Denver. How long will it take for the car to get to Denver?

A. 40 minutes

B. 60 minutes

C. 4 hours

D. 5 hours

E. 6 hours

6. Use the diagram below to answer the question that follows.

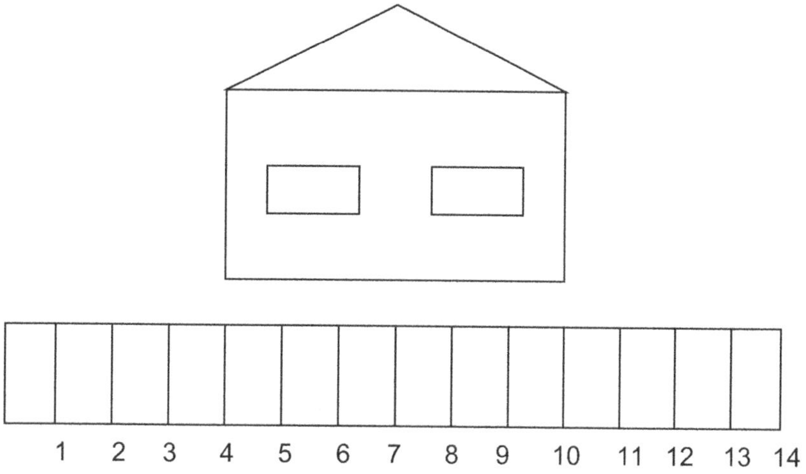

If each rectangle in the ruler below the picture of the house is one unit and the actual length of the house is 36 feet, then what is the scale of the diagram of the house?

A. 1 unit = 6 feet

B. 1 unit = 7.2 feet

C. 1 unit = 9 feet

D. 1 unit = 12 feet

E. 1 unit = 36 feet

7. What is the most appropriate unit of measure for determining the dimensions of a sofa?

A. feet and inches

B. pounds

C. yards

D. ounces

E. tons

8. What is the remainder when 11 is divided by 3?

A. .06

B. .27

C. 2

D. 3

E. 9

9. Mrs. Ramirez is inviting 12 children to her son's birthday party. The children will play pin the tail on the donkey. Mrs. Ramirez has already made 40 tails for the game. She wants to give each child 4 tails to play the game. How many more tails does she need to make?

A. 4

B. 8

C. 10

D. 12

E. 28

10. Yesterday the temperature was 90 degrees. Today it is 10% cooler than yesterday. What is today's temperature?

A. 80 degrees

B. 81 degrees

C. 91 degrees

D. 99 degrees

E. 100 degrees

11. A class contains 20 students. On Tuesday 5% of the students were absent. On Wednesday 20% of the students were absent. How many more students were absent on Wednesday than on Tuesday?

A. 1

B. 2

C. 3

D. 4

E. 5

12. $^{1}/_{3} - ^{1}/_{7} = ?$

A. $^{1}/_{21}$

B. $^{1}/_{4}$

C. $^{3}/_{7}$

D. $^{4}/_{21}$

E. $-^{1}/_{4}$

13. Mark's final grade for a course is based on the scores from two tests, A and B. The score from test A counts toward 35% of his final grade. The score from test B counts toward 65% of his final grade. What equation is used to calculate Mark's final grade for this course?

A. (.65A + .35B)

B. (.35A + .65B)

C. (.35A + .65B) ÷ 2

D. A + B

E. (A + B) ÷ 2

14. If A represents the number of apples purchased at 20 cents each and B represents the number of bananas purchased at 25 cents each, what equation represents the total value of the purchase?

A. .45AB

B. .45 ÷ AB

C. .20A + .25B

D. .25A + .20B

E. .45 + (A × B)

15. If $x - 3 + 5x = 33$, then $x = ?$

A. 6

B. 7

C. 8

D. 9

E. 12

16. Pat wants to put wooden trim around the floor of her family room. Each piece of wood is 1 foot in length. The room is rectangular and is

12 feet long and 10 feet wide. How many pieces of wood does Pat need for the entire perimeter of the room?

A. 22

B. 44

C. 100

D. 120

E. 144

17. Ben uses one bag of dog food every 6 days to feed his dog. Approximately how many bags of dog food would Ben require for two months?

A. 5

B. 6

C. 9

D. 10

E. 20

18. The snowfall for November is 5 inches less than for December. If the total snowfall for November and December is 35 inches, what was the snowfall for November?

A. 10 inches

B. 15 inches

C. 20 inches

D. 30 inches

E. 40 inches

19. Records indicate that there were 12 hospitals in Johnson County in 1995, but this number had increased to 15 hospitals in 2014. There were 12 births on average per hospital in Johnson County in 1995.

The total number of births in Johnson County was 240 in 2014. By what amount does the average number of births per hospital in Johnson County for 2014 exceed those for 1995?

A. 3 births per hospital

B. 4 births per hospital

C. 15 births per hospital

D. 16 births per hospital

E. 96 births per hospital

20. A magician has a bag of colored scarves for a magic trick that he performs. The bag contains 3 blue scarves, 1 red scarf, 4 green scarves, and 2 orange scarves. If the magician removes scarves at random and the first scarf she removes is red, what is the probability that the next scarf will be orange?

A. $^{1}/_{2}$

B. $^{2}/_{7}$

C. $^{1}/_{9}$

D. $^{2}/_{9}$

E. $^{2}/_{10}$

21. Marta can walk one mile in 17 minutes. At this rate, how long would it take her to walk 5 miles?

A. 1 hour and 5 minutes

B. 1 hour and 7 minutes

C. 1 hour and 8 minutes

D. 1 hour and 15 minutes

E. 1 hour and 25 minutes

22. Simplify the following expression: $-117 + (-25) + 45$

A. –47

B. –97

C. –137

D. 137

E. 187

23. **Use the information below to answer the question that follows.**

Appleton	Brownsville	Charlestown	Durham	Easton
687 feet below sea level	1586 feet above sea level	253 feet below sea level	542 feet below sea level	1621 feet above sea level

As part of a geography class, students are required to learn the distance above and below sea level of certain towns in their area. What was the difference in feet between the highest and lowest towns in their area according to the above table?

A. 66 feet

B. 621 feet

C. 874 feet

D. 2273 feet

E. 2308 feet

24. Clark County had 135,298 cases of infectious disease last year, while Davidson County had 207,121 cases. What number is the best estimate of how many more cases of infectious disease there were in Davidson County?

A. 12,000

B. 62,000

C. 70,000

D. 71,000

E. 72,000

25. What is the best estimate of $5,012 \times 12$?

A. 50,000

B. 52,000

C. 60,000

D. 70,000

E. 600,000

26. **Please use the diagram below to answer the question that follows.**

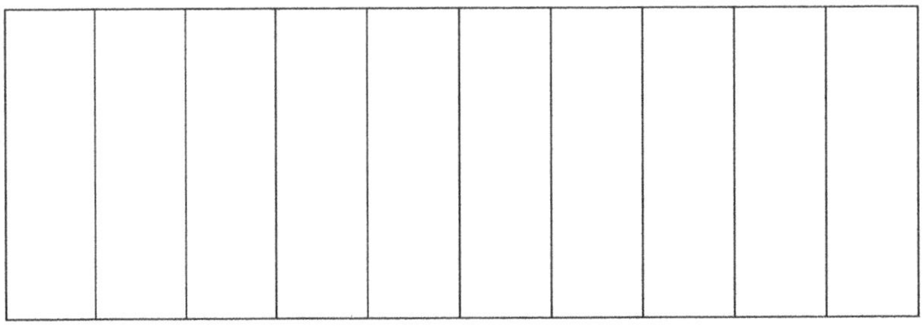

The above diagram depicts a football field. The field is 30 yards wide and 100 yards long. Paint is sprayed on the field in lines that are 10 yards apart, as indicated by the vertical lines in the diagram above. Paint is also sprayed around the entire perimeter of the field. In total, how many yards of paint are sprayed onto the field?

A. 290

B. 300

C. 530

D. 560

E. 590

27. Solve for x: $3x - 4 - x = 12$

A. –2

B. –4

C. 2

D. 4

E. 8

28. A martial arts class has 53 students at the beginning of the year. 15 students have black belts, 22 have brown belts, 8 have blue belts, and 8 have belts of other colors. By the end of the year, 3 of the students with brown belts and 2 of the students with belts of other colors have dropped out of the class. In addition, 4 new students have joined the class.

Which of the following facts can be determined from the information above?

A. The total number of students in the class.

B. The number of students in the class with brown belts.

C. The number of students in the class with blue belts.

D. The number of students in the class with black belts.

E. The number of students in the class with belts of other colors.

29. Sam's final grade for a class is based on his scores from a midterm test (M), a project (P), and a final exam (F). The midterm test counts twice as much as the project, and the final exam counts twice as

much as the midterm. Which mathematical expression below can be used to calculate Sam's final grade?

A. P + M + F

B. P + M + 2F

C. P + 2M + F

D. P + 2M + 2F

E. P + 2M + 4F

30. Bart is riding his bike at a rate of 12 miles per hour. He arrives in the town of Wilmington at 3:00 pm. The town of Mount Pleasant is 50 miles from Wilmington. How far will Bart be from Mount Pleasant at 5:00 pm if he continues riding his bike at this speed?

A. 12 miles

B. 20 miles

C. 24 miles

D. 26 miles

E. 36 miles

31. A ticket office sold 360 more tickets on Friday than it did on Saturday. If the office sold 2570 tickets in total during Friday and Saturday, how many tickets did it sell on Friday?

A. 360

B. 1105

C. 1465

D. 1565

E. 2210

32. **Use the chart below to answer the question that follows.**

X	*Y*
2	4
4	16
6	
8	64
10	100

The chart above shows the mathematical relationship between *X* and *Y*. What value of *Y* is missing from the chart?

A. 24

B. 30

C. 32

D. 36

E. 48

33. Tom's height increased by 10% this year. If Tom was 5 feet tall at the beginning of the year, how tall is he now?

A. 5 feet 1 inch

B. 5 feet 5 inches

C. 5 feet 6 inches

D. 5 feet 10 inches

E. 6 feet

34. Read the problem below and answer the question that follows.

Mary left home at 10:00 am. She drove 150 miles to Newton at a constant rate of 70 miles per hour. She then rested for 30 minutes before driving 105 miles to Lordville. What time was it when she arrived in Lordville?

What piece of information is needed in order to solve the problem?

A. The number of gallons of gas she used for the journey.

B. The speed that Mary traveled from Newton to Lordville.

C. The speed limit on the road from Newton to Lordville.

D. The amount of time she rested in Lordville.

E. Her arrival time at her house on her return journey.

35. Read the problem below and answer the question that follows.

> Dan rode his horse 2 miles to his neighbor's house. It took the horse 15 minutes to make this journey. From his neighbor's house, Dan rode his horse 3 miles into town. What is the average pace of Dan's horse in miles per hour for these two journeys?

What piece of information is needed in order to solve the problem?

A. The distance from Dan's house into town.

B. The amount of time Dan stayed at his neighbor's house.

C. The length of the stride of Dan's horse.

D. Whether Dan's horse trotted or galloped.

E. The amount of time it took to go from the neighbor's house into town.

36. Which mathematical expression is equivalent to 2H + 3H?

A. 5H

B. 5 + H

C. 6H

D. H^5

E. H^6

37. Which of the following numbers is between 2,368,741 and 2,654,802?

A. 2,281,414

B. 2,306,549

C. 2,367,988

D. 2,683,699

E. 2,645,972

38. Beth needs to calculate 15% of 60. Which equation below can she use in order to do so?

A. 60 × 15

B. 60 × 1.5

C. 60 × .15

D. 60 × .0015

E. 60 ÷ 15

39. Which of the following mathematical statements is correct?

A. $.001 < .0001 < .00001$

B. $.001 < .0010 < .00100$

C. $.0001 < .001 < .01$

D. $.0010 < .001 < .01$

E. $.0100 < .010 < .10$

40. $^1/_3 > ? > {}^1/_9$

A. $^1/_2$

B. $^1/_4$

C. $^1/_{10}$

D. $^2/_3$

E. $^2/_5$

41. Carlos buys 2 pairs of jeans for $22.98 each. He later decides to exchange both pairs of jeans for 3 sweaters which cost $15.50 each. Which equation can Carlos use to calculate the extra money he will have to pay for the exchange?

A. $2 \times (22.98 - 15.50)$

B. $3 \times (22.98 - 15.50)$

C. $(3 \times 22.98) - (2 \times 15.50)$

D. $(3 \times 15.50) - (2 \times 22.98)$

E. $(3 \times 15.50) + (2 \times 22.98)$

42. If the value of *x* is less than .06 but greater than .006, which one of the following could be *x*?

A. .05

B. .005

C. .0005

D. .0006

E. .00005

43. A museum counts its visitors each day and rounds each daily figure up or down to the nearest 5 people. 104 people visit the museum on Monday, 86 people visit the museum on Tuesday, and 81 people visit the museum on Wednesday. Which figure below best represents the amount of visitors to the museum for the three days, after rounding?

A. 260

B. 265

C. 270

D. 275

E. 280

44. Jason does the high jump for his high school track and field team. His first jump is at 3.246 meters. His second is 3.331 meters, and his third is 3.328 meters. If the height of each jump is rounded to the nearest one-hundredth of a meter (also called a centimeter), what is the estimate of the total height for all three jumps combined?

A. 9.80

B. 9.89

C. 9.90

D. 9.91

E. 10.00

45. Use the information below to answer the question that follows.

- The baseball team practices every Tuesday and Friday.
- There will be no practice during the last full week of the month.
- There will be no practice in the event of rain.

If there is practice today, which of the following conclusions can be made?

A. It is the last full week of the month.

B. It is a Tuesday or it is not raining.

C. It is a Tuesday and it is raining.

D. It is a Tuesday or a Friday.

E. It is the last full week of the month or it is raining.

46. Use the table below to answer the question that follows.

Regional Railway Train Service	
Departure Time	Arrival Time
9:50 am	10:36 am
11:15 am	12:01 pm
12:30 pm	1:16 pm
2:15 pm	3:01 pm
?	5:51 pm

The journey on the Regional Railway is always exactly the same duration.

What is the missing time in the chart above?

A. 3:30 pm

B. 4:15 pm

C. 4:30 pm

D. 5:05 pm

E. 5:15 pm

47. Use the chart below to answer the question that follows.

United Kingdom and Entire World Coal Consumption in Tons		
Year	**United Kingdom**	**Entire World**
1925	3528	8741
1945	3679	9523
1965	3598	7413
1985	2565	6528
2005	1201	4889
2010	800	4665

How many tons of coal were consumed outside the United Kingdom in 1965?

A. 3815

B. 3915

C. 4815

D. 6528

E. 7413

48. Use the chart below to answer the question that follows.

Electricity and Gas Consumption for the Past Ten Years

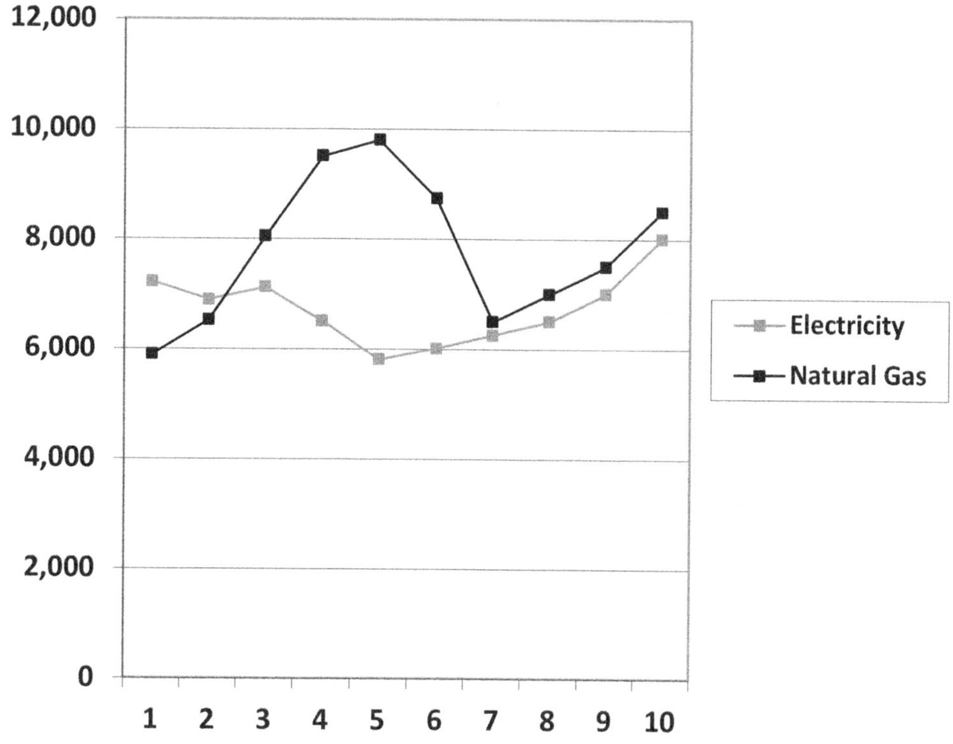

According to the graph, the greatest annual difference in units between the consumption of electricity and the consumption of natural gas occurred in which year?

A. 2,000 units

B. 3,000 units

C. 4,000 units

D. 5,000 units

E. 16,000 units

49. Use the chart below to answer the question that follows.

Crop Production

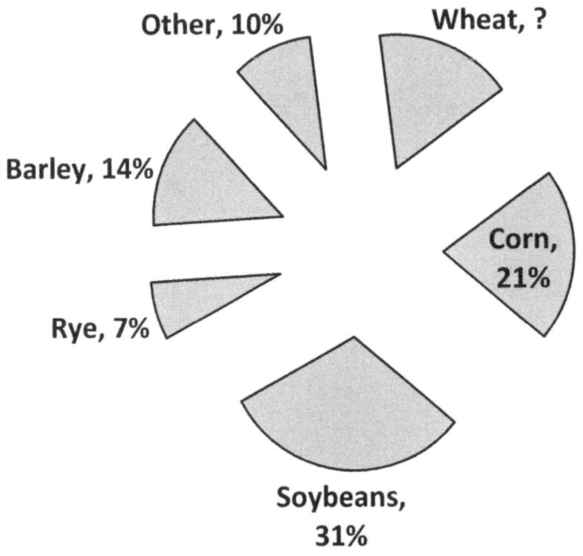

The percentage for wheat production is not provided in the above the chart. What is the percentage of wheat production?

A. 9%

B. 12%

C. 13%

D. 17%

E. 29%

50. Use the graph below to answer the question that follows.

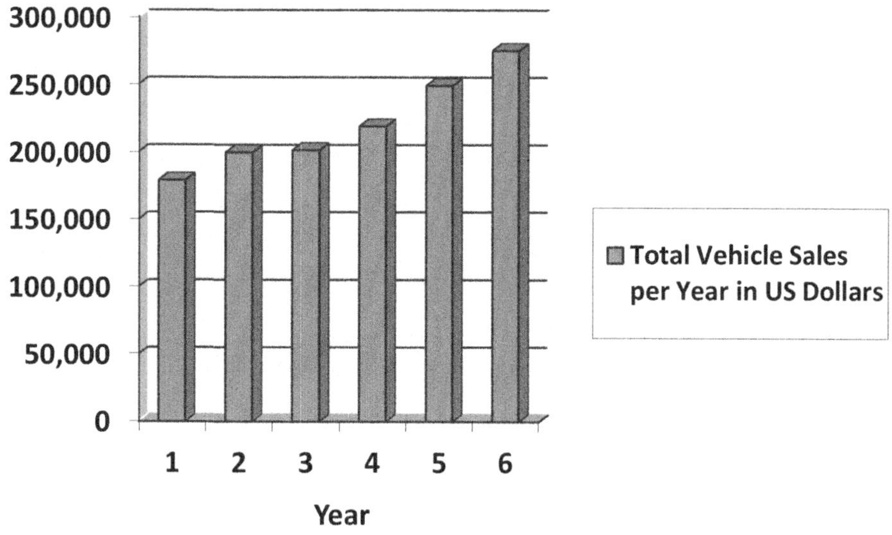

The graph shows sales of vehicles during a six-year period. Between which years did the sales increase the most?

A. Years 1 and 2

B. Years 2 and 3

C. Years 3 and 4

D. Years 4 and 5

E. Years 5 and 6

CBEST Math Practice Test 2 – Answers

1. The correct answer is E.

We have both fractions and decimals in this problem.

Convert the fractions in the mixed numbers to decimals.

$^3/_4 = 3 \div 4 = 0.75$

$^1/_5 = 1 \div 5 = 0.20$

Then represent the mixed numbers as decimal numbers.

Person 1: $14^3/_4 = 14.75$

Person 2: $20^1/_5 = 20.20$

Person 3: 36.35

Then add all three amounts together to find the total.

14.75 + 20.20 + 36.35 = 71.30

2. The correct answer is B.

First, divide the total price for the multi-purchase by the number of items.

In this case, \$150 ÷ 5 = \$30 for each of the five books.

Then, subtract this amount from the original price to get your answer.

\$40 – \$30 = \$10

Alternatively, you can use the method explained below.

Calculate the total price for the five books without the discount.

5 × \$40 = \$200

Then subtract the discounted price of \$150 from the total.

\$200 - \$150 = \$50

Then divide the total savings by the number of books to determine the savings on each book.

\$50 total savings ÷ 5 books = \$10 savings per book

3. The correct answer is B.

You need to find the total points for all the females by multiplying their average by the number of female students. Then do the same to find the total points for all the males.

Females: 60 × 95 = 5700

Males: 60 × 90 = 5400

Then add these two amounts together to get the total for the group.

5700 + 5400 = 11,100

Then divide by the total number of students in the class to get your solution.

11,100 ÷ 120 = 92.5

So, the correct average is 92.5

4. The correct answer is C.

Find the dollar amount of the discount first.

$15 original price – $12 sales price = $3 discount

Then divide the discount into the original price to get the percentage.

$3 ÷ $15 = 0.20 = 20%

5. The correct answer is C.

Divide the miles per hour into the distance left in order to get the time needed.

240 miles remaining ÷ 60 mph = 4 hours left to travel

6. The correct answer is A.

Count the number of units that the house spans, rather than trying to subtract units from the total of 14.

If we count the number of units below the house in the drawing, we can see that the house spans 6 units.

Divide this result into the actual length of the house (36 feet) to get the scale of the drawing.

36 feet ÷ 6 units = 6 feet represented by each unit

7. The correct answer is A.

Pounds, ounces, and tons are used to measure the weight of items, not their dimensions or linear measurements. Feet, inches, and yards are used to measure linear dimensions. Feet and inches are suitable for small items which one usually finds indoors, while yards are used for larger items which one usually finds outdoors. A sofa is an item of furniture which is used inside the house, so "feet and inches" is the best answer.

8. The correct answer is C.

The remainder is the amount that is left over after you divide a problem into whole numbers. These whole numbers are referred to as factors. So, ask yourself what numbers can be calculated by multiplying by 3.

$1 \times 3 = 3$

$2 \times 3 = 6$

$3 \times 3 = 9$

$4 \times 3 = 12$

12 is greater than 11, so the nearest product to 11 from the list above is 9. Finally, we subtract these two numbers to get the remainder.

$11 - 9 = 2$

9. The correct answer is B.

If there are 12 children and each one is supposed to receive 4 items, we can do the calculation as follows:

12 children × 4 items per child = 48 items required in total

Now subtract the total from the amount she already has in order to determine how many more she needs.

48 items required in total – 40 items available = 8 items still needed

10. The correct answer is B.

First, you need to determine the difference in degrees.

90 degrees yesterday × 10% = 9 degrees cooler today

Then subtract to get your answer.

90 degrees yesterday – 9 degrees cooler today = 81 degrees today

11. The correct answer is C.

First of all, you have to find out how many students were absent on Tuesday. To find the number of absent students, you have to multiply the total number of students in the class by the percentage of the absence for Tuesday.

20 students in total × 5% = 1 student absent on Tuesday

Now calculate the absences for Wednesday in the same way.

20 students in total × 20% = 4 students absent on Wednesday

The problem is asking you how many more students were absent on Wednesday than Tuesday, so you need to subtract the two figures that you have just calculated above.

4 students absent on Wednesday – 1 student absent on Tuesday = 3 students

So, 3 more students were absent on Wednesday.

12. The correct answer is D.

Remember that when you see fractions that have different numbers on the bottom, you have to find the lowest common denominator (LCD).

In this problem, our LCD is 21.

So, we have to convert to fractions into the LCD. To calculate the LCD, you have to multiply the numerator and denominator by the same number.

$$\frac{1}{3}-\frac{1}{7}=$$

$$\left(\frac{1}{3}\times\frac{7}{7}\right)-\left(\frac{1}{7}\times\frac{3}{3}\right)=$$

$$\frac{7}{21}-\frac{3}{21}=\frac{4}{21}$$

13. The correct answer is B.

The two tests are being given different percentages, so each test needs to have its own variable.

A for test A and B for test B.

Since A counts for 35% of the final grade, we set 35% to a decimal and put the decimal in front of the variable so that the variable will have the correct weight.

So, the value of test A is .35A

Test B counts for 65%, so the value of test B is .65B

The final grade is the sum of the values for the two tests.

So, we add the above products together to get our equation.

.35A + .65B

14. The correct answer is C.

Here is another question on setting up equations. The quantity of apples is represented by A and the quantity of bananas is represented by B.

Apples are 20 cents each, while bananas cost 25 cents.

So, the cost of apples is .20A

The cost of bananas is .25B

To get the total cost, we have to add the cost of the apples to the cost of the bananas

.20A + .25B

15. The correct answer is A.

Remember that for problems like this one, you need to get the terms that have a variable on one side of the equation. On the other side of the equation, you should have the whole numbers.

STEP 1: Our first step is to deal with the whole numbers.

$x - 3 + 5x = 33$

$x - 3 + 3 + 5x = 33 + 3$

$x + 5x = 36$

STEP 2: Then simplify the numbers that have variables.

$x + 5x = 36$

$6x = 36$

STEP 3: To solve the problem, remove the number in front of the variable by dividing.

$6x \div 6 = 36 \div 6$

$x = 6$

16. The correct answer is B.

Remember that the perimeter is the measurement along the outside edges of the rectangle or other area.

The formula for perimeter is as follows:

P = 2W + 2L

If the room is 12 feet by 10 feet, we need 12 feet × 2 feet to finish the long sides of the room and 10 feet × 2 feet to finish the shorter sides of the room.

$(2 \times 10) + (2 \times 12) =$

$20 + 24 = 44$

17. The correct answer is D.

Most months have 30 or 31 days. In this problem, we are being asked to do a calculation for a 2-month period, so we are dealing with 60 to 62 days.

For the purposes of estimation, we can use 60 days.

Ben uses a bag of dog food every 6 days.

So, we divide the total period by the number of days to get the required amount.

60 day period ÷ 6 days each bag = 10 bags needed for 60 days

18. The correct answer is B.

You will notice in this problem that we are dealing with two months, November and December.

STEP 1: Look at the total.

The total for the two months is 35 inches.

STEP 2: Determine whether one month is higher than the other.

In this problem, one month has 5 inches more than the other, so you have to subtract the difference first of all.

$35 - 5 = 30$

STEP 3: Now divide this amount by two to allocate each part to the two months.

$30 \div 2 = 15$

STEP 4: Look again at the problem to see if you are calculating the amount for the high month or the low month.

Here, the amount for November is lower than the amount for December.

So, we know that November had 15 inches of snowfall.

If the problem had asked for the higher month, you would then need to add back the difference.

So, December's snowfall is 15 + 5 = 20

STEP 5: Check your result by adding the amounts for the two months together.

15 + 20 = 35

19. The correct answer is B.

The problem is asking you for the amount that the average number of births per hospital in Johnson County for 2014 exceeded those for 1995.

STEP 1: First we have to calculate the average for 2014.

In order to calculate an average, you have to divide the total amount by the number of items in each data set.

For 2014, we have 240 total births and 15 hospitals in the data set.

240 ÷ 15 = 16 births on average per hospital for 2014

STEP 2: Now calculate the average for 1995

In our problem, this average is provided.

We can see that there were 12 births on average per hospital in Johnson County in 1995.

STEP 3: Now subtract the averages for the two years to get your answer.

16 – 12 = 4 more births per hospital in 2014

20. The correct answer is D.

This is a question on probability, which is a statistical measure.

STEP 1: Determine the total amount in the data set before any items are removed.

Here, we have a bag that contains 3 blue scarves, 1 red scarf, 4 green scarves, and 2 orange scarves.

3 + 1 + 4 + 2 = 10 items in the data set

STEP 2: Determine the numbers of items in the data set after items have been removed.

One scarf is removed.

10 – 1 = 9 items left in the data set

STEP 3: Determine the amount in the subset.

The problem is asking for the orange scarf subset. So, we have 2 orange scarves in the subset.

Note that if the problem were asking you for the red scarf subset, you would have to subtract the item that has already been removed from the subset.

STEP 4: The probability is expressed as a fraction. The amount in the subset (2 orange scarves) goes on the top of the fraction and the amount of items left in the data set (9 items left) goes on the bottom.

So, the answer is $^2/_9$.

21. The correct answer is E.

This is another measurement problem.

STEP 1: You need to multiply the number of miles that she is going to travel by the amount of time it takes her to travel one mile.

17 minutes for 1 mile × 5 miles to travel = 85 minutes needed

STEP 2: Now express the result in hours an minutes, remembering of course that an hour has 60 minutes.

85 minutes – 60 minutes = 25 minutes left

So, the answer is 1 hour and 25 minutes.

22. The correct answer is B.

Remember to be careful with the negatives.

STEP 1: Deal with the negative numbers first.

–117 + (–25) + 45 =

–117 – 25 + 45 =

(–117 – 25) + 45 =

–142 + 45

STEP 2: Then deal with the positive number.

–142 + 45 =

–97

23. The correct answer is E.

STEP 1: Look at the chart to see which town is the highest.

Here, Eaton is the highest at 1621 feet above sea level.

STEP 2: Look at the chart to see which town is the lowest.

Appleton is the lowest at 687 feet below sea level.

STEP 3: Now add these two amounts together to find the total distance between the high point and the low point.

1621 + 687 = 2308

Note that if both points are above ground, you need to subtract the two amounts.

24. The correct answer is E.

The problem tells us that Clark County had 135,298 cases of infectious disease last year, while Davidson County had 207,121 cases.

STEP 1: Round each number up or down to the nearest thousand.

207,121 is rounded down to 207,000.

135,298 is rounded down to 135,000.

STEP 2: Subtract the two figures to estimate the difference.

207,000 – 135,000 = 72,000

25. The correct answer is C.

To get best estimate of 5,012 $\times$ 12, you need to round only the larger number up or down.

5,012 is rounded to 5,000

12 is not rounded in this case since the problem is asking for the *best* estimate.

Then multiply.

5,000 × 12 = 60,000

26. The correct answer is C.

This is a linear measurement problem.

STEP 1: Calculate the perimeter.

The field is 30 yards wide and 100 yards long.

The formula for perimeter is as follows:

P = 2W + 2L

(2 × 30) + (2 × 100) =

60 + 200 = 260 yards for the perimeter

STEP 2: Determine the linear total of all of the lines on the interior of the field. Be careful not to count the ends of the field again.

By counting the lines on the diagram, we can see that we have 9 lines on the interior of the field.

Each line will be 30 yards in distance, since that is the width of the field.

9 lines × 30 yards each = 270 yards for the lines

STEP 3: Now add these two amounts together to get your answer.

260 + 270 = 530

27. The correct answer is E.

Remember to deal with the whole number first.

$3x - 4 - x = 12$

$3x (- 4 + 4) - x = 12 + 4$

$3x - x = 16$

Then deal with the variable.

$3x - x = 16$

$2x = 16$

Then divide to get your answer.

$2x \div 2 = 16 \div 2$

$x = 8$

28. The correct answer is A.

We cannot calculate the number of students in the class with belts of particular colors because we do not know the color of the belts the new students.

The problem is telling us how many students there are in each group and how many of each group have left.

The problem also tells us how many students in total have joined, so we can calculate the new total number of students.

29. The correct answer is E.

Sam's final grade for a class is based on his scores from a midterm test (M), a project (P), and a final exam (F), but the midterm test counts twice as much as the project, and the final exam counts twice as much as the midterm. Therefore, we have to count variable M twice.

The value of the midterm is doubled and variable F is double of the midterm, so we have to count variable F 4 times.

So, the equation is: P + 2M + 4F

30. The correct answer is D.

The problem tells us that Bart rides at a rate of 12 miles per hour. We also know that he arrives in the town of Wilmington at 3:00 pm. The question is asking us how far Bart will be from Mount Pleasant at 5:00 pm.

STEP 1: Calculate the time difference.

5:00 pm – 3:00 pm = 2 hours difference

STEP 2: Calculate the distance traveled.

12 miles per hour × 2 hours = 24 miles traveled

STEP 3: Calculate the distance left.

The town of Mount Pleasant is 50 miles from Wilmington.

50 miles to travel – 24 miles traveled = 26 miles left

31. The correct answer is C.

The ticket office sold 360 more tickets on Friday than it did on Saturday. The office sold 2570 tickets in total during Friday and Saturday.

STEP 1: Subtract the excess.

2570 – 360 = 2210

STEP 2: Allocate the above figure to each day.

2210 ÷ 2 = 1105

STEP 3: Calculate Friday's amount by adding back in the excess.

1105 + 360 = 1465

32. The correct answer is D.

You need to find the relationships between the numbers provided in the chart in order to determine the missing value.

STEP 1: Consider whether a relationship between the numbers on the first row of the table can be found based on addition or subtraction.

Look at each of the sets of numbers on a line by line basis.

On the first line, we have 2 in the left column and 4 in the right column.

So, we can get to the value in the left column by adding 2.

STEP 2: Try out the value calculated in step 1 for the next row of numbers.

$4 + 2 \neq 16$

STEP 3: If the relationship does not work for the second row of number we have to consider whether the relationship between the numbers is based on multiplication or division.

Returning to row 1 of the table, we can determine that: $2 \times 2 = 4$

STEP 4: Try this operation on the second row of numbers.

$4 \times 2 \neq 16$

STEP 5: Try to determine if any other relationship is possible.

Since $2 \times 2 = 4$ on the first row of the table, we can also try multiplying each subsequent number by itself.

STEP 6: Try this new relationship for the second and subsequent rows.

Row 2: $4 \times 4 = 16$

Row 4: $8 \times 8 = 64$

Row 5: $10 \times 10 = 100$

STEP 7: Calculate the value missing from row 3.

Row 3: $6 \times 6 = 36$

33. The correct answer is C.

Tom was 5 feet tall at the beginning of the year, and his height increased by 10% this year.

STEP 1: Calculate the beginning height in inches. Remember that there are 12 inches in a foot.

5 feet × 12 inches per foot = 60 inches in height

STEP 2: Calculate the increase in height.

60 inches × 10% = 6 inches

STEP 3: Calculate the new height by adding the increase to the number at the beginning.

5 feet + 6 inches = 5 feet 6 inches

34. The correct answer is B.

The problem is asking us what information is required in order to determine the time Mary arrived in Lordville.

STEP 1: In order to determine the arrival time at a new destination, we need to know the time the person began the journey and the amount of time he or she traveled.

From the problem, we know that Mary left home at 10:00 am.

STEP 2: In order to know the amount of time a person travels, we need to know the amount of miles traveled and the speed.

The problem tells us that she drove 150 miles to Newton at a constant rate of 70 miles per hour.

However, her journey is ending in Lordville, so we also need to know the amount of miles traveled and the speed of traveling to Lordville.

From the problem, we know that she drove 105 miles to Lordville after resting in Newton.

STEP 3: Determine which information is missing.

We do not know the speed she traveled to Lordville, so B is the correct answer.

35. The correct answer is E.

This question is similar to the previous one.

To calculate the average pace or speed, we need to know the speed for each journey.

You will recall from the previous problem that in order to calculate the speed of travel, we need to know the distance traveled and the amount of time for the journey.

The problem tells us that Dan rode his horse 2 miles to his neighbor's house and that it took 15 minutes for this journey.

So, we have both the distance traveled and the amount of time for the first journey.

The problem also states that Dan made a second journey, riding his horse 3 miles into town from his neighbor's house.

So, we have the distance traveled for the second journey, but we do not have the amount of time for the second journey.

36. The correct answer is A.

For addition problems like this one, remember that you can just add the numbers in front of the variables if both terms have the same variable.

2H + 3H = 5H

37. The correct answer is E.

If you have problems like this one, you should line all of the numbers up in a column to help you find the relationship. From the facts of the problem, we have 2,368,741 and 2,654,802. If you are unable to solve the problem visually, put the other numbers from the answer choices in between these two numbers to determine your answer.

<u>Answer A:</u>
2,368,741
2,281,414
2,654,802

2,281,414 is less than 2,368,741, so answer A is not the correct choice.

Answer B:
2,368,741
2,306,549
2,654,802

2,306,549 is less than 2,368,741, so answer B is not the correct choice.

Answer C:
2,368,741
2,367,988
2,654,802

2,367,988 is less than 2,368,741, so answer C is not the correct choice.

Answer D:
2,368,741
2,683,699
2,654,802

2,683,699 is greater than 2,654,802, so answer D is not the correct choice.

Answer E:
2,368,741
2,645,972
2,654,802

From the three figures above, we can see that answer E is correct since 2,645,972 is greater than 2,368,741 and less than 2,654,802.

38. The correct answer is C.

In order to calculate the percent of a number, we multiply the percent by the number.

Remember to convert the percent to a decimal by placing a decimal point two places from the right.

15% × 60 =

.15 × 60 =

60 × .15

39. The correct answer is C.

This question is similar to question 37 above.

Remember to place the numbers into column. For numbers that are decimals like in this problem, you can add zeroes to help you.

<u>Answer A:</u>
.00100
.00010
.00001

.00100 is greater than .00010, but the problem stipulates that .00100 is less than .00010, so answer A is not correct.

<u>Answer B:</u>
.00100
.00100
.00100

All of the figures provided in answer B are equivalent to one another, so it is not the correct answer.

<u>Answer C:</u>
.0001
.0010
.0100

.0001 is less than .0010, and .0010 is less than .0100, so answer C is correct.

Answer D:
.0010
.0010
.0100

Answer E:
.0100
.0100
.1000

The first two figures provided in answer D and in answer E are equivalent to one another, so they are not the correct answers.

40. The correct answer is B.

If all of the numerators are 1, we know that the denominator of the unknown faction must lie between the denominators of the fractions stated in the problem.

The problem asks: $^1/_3 > ? > {}^1/_9$

The denominator 3 from the equation above is greater than the denominator of 4 from answer B. The denominator of 4 from answer B is less than the denominator of 9 from the equation above. So, B is the correct answer.

Note that if you are not able to find the correct answer from the options that have 1 in the numerator, you will need to calculate the lowest common denominator for all of the fractions.

41. The correct answer is D.

The problem tells us that Carlos buys 2 pairs of jeans for $22.98 each, and then he decides to exchange both pairs of jeans for 3 sweaters which cost $15.50 each.

STEP 1: Calculate the amount of money spent on the original purchase of the jeans.

2 × $22.98 = $45.96

STEP 2: Calculate the value of the items acquired in the exchange, which in this case, is the value of the sweaters.

3 × $15.50 = $46.50

STEP 3:

Calculate the difference between the value of the items acquired and the amount of money originally spent.

Value of the items acquired

3 × $15.50 = $46.50

Amount of money originally spent

2 × $22.98 = $45.96

Difference:

(3 × $15.50) – (2 × $22.98)

42. The correct answer is A.

Remember to put the numbers in columns if you are still not used to determining the value by visual inspection.

Here, we see that .05 is less than .06, and that .050 is also greater than .006, so answer A is the correct choice.

43. The correct answer is C.

If you look at the answer choices, you will see that they are given in the nearest increments of 5.

So, we have to round the figures stated in the problem up or down to the nearest increment of 5.

104 on Monday is rounded to 105.

86 on Tuesday is rounded down to 85.

81 is rounded down to 80.

Then add these three figures together to get your result.

105 + 85 + 80 = 270

44. The correct answer is D.

We know that we have to round to the nearest hundredth.

The hundredth decimal place is the number 2 positions to the right of the decimal.

For example, .01 is 1 one hundredth.

In our question, the first jump of 3.246 is rounded up to 3.25

The second jump of 3.331 is rounded down to 3.33

The third jump of 3.328 is rounded up to 3.33

Then add these three figures together to get your answer.

3.25 + 3.33 + 3.33 = 9.91

45. The correct answer is D.

If there is practice today, we can conclude that it is a Tuesday or a Friday. The facts tell us that there will be no practice during the last full week of the month and that there will be no practice in the event of rain.

46. The correct answer is D.

You have to find the relationship between the number given in each row in the left column and the corresponding number in the right column. "9:50 am to 10:36 am" represents a journey time of 46 minutes.

11:15 to 12:01 is also 46 minutes, and so on.

If we go 46 minutes back from 5:51 pm, we arrive at 5:05 pm.

47. The correct answer is A.

Look at the row of the table that displays information for the year 1965, which is the third row.

We want to look at the last column of row 3 to find the figure for the entire world. The figure provided is 7413 tons.

Next, we need to subtract the amount for the United Kingdom from the amount for the entire world in order to calculate the consumption outside the United Kingdom.

The United Kingdom amount for 1965 was 3598 tons.

7413 – 3598 = 3815 tons consumed outside the United Kingdom in 1965

48. The correct answer is C.

For line graphs, look at the distance between the lines.

Here, we can see that the biggest difference, in other words, the largest gap between the two lines, is in Year 5.

Natural gas consumption was nearly 10,000 units in Year 5, while electricity consumption was approximately 6,000 units for the same year.

To calculate the difference, you need to subtract. 10,000 – 6,000 = 4,000

49. The correct answer is D.

First, you need to add the stated percentages together to get the amount that is represented by the given slices of the pie chart.

21% + 31 % + 7% + 14% + 10% = 83%

Then subtract from 100% to get your answer.

100% - 83% = 17% wheat production

50. The correct answer is D.

For bar graphs, you will need to compare the height of the bars to one another.

For Years 1 to 4, there is not much change from year to year.

We can see from looking at the graph that the biggest growth was between years 4 and 5 or years 5 and 6.

However, the difference between the bars for year 5 and year 6 is slightly less than the difference between the bars for year 4 and year 5.

So, we have to choose answer D.

CBEST Practice Math Test 3

1. Captain Smith needs to purchase rope for his fleet of yachts. He owns 26 yachts and needs 6 feet 10 inches of rope for each one. How much rope does he need in total?

   A. 152 feet

   B. 177 feet 8 inches

   C. 257 feet 8 inches

   D. 260 feet

   E. 412 feet

2. Use the diagram below to answer the question that follows.

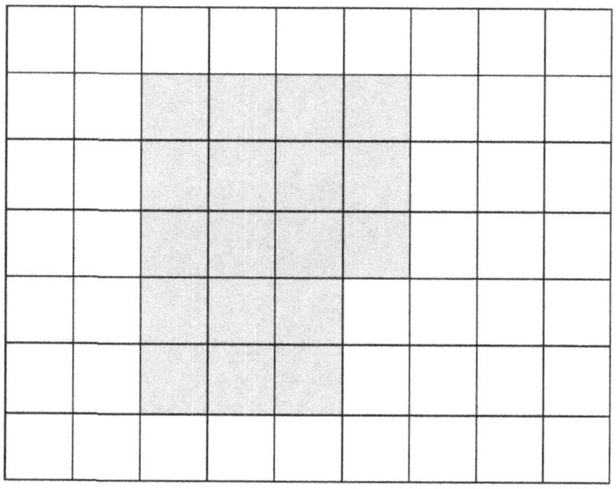

   Each square in the diagram above is one foot wide and one foot long. The gray area of the diagram represents the layout of New Town's water reservoir. What is the perimeter in feet of the reservoir?

   A. 16 feet

   B. 17 feet

C. 18 feet

D. 20 feet

E. 32 feet

3. During this term, Tom got the following scores on his math tests: 98, 78, 89, 85, and 90. What is the average of Tom's scores?

A. 78

B. 85

C. 88

D. 89

E. 98

4. What is the best unit of measure for expressing the weight of a bag of sugar?

A. ounces

B. inches

C. pints

D. quarts

E. tons

5. Use the table below to answer the question that follows.

Part	Total Number of Questions	Number of Questions Answered Correctly
1	15	12
2	25	20
3	35	32
4	45	32

Chantelle took a test that had four parts. The total number of questions on each part is given in the table above, as is the number of questions Chantelle answered correctly. What was Chantelle's percentage of correct answers for the entire test?

A. 75%

B. 80%

C. 86%

D. 90%

E. 96%

6. Linda uses two bottles of ink every 5 days for her graphic design business. Approximately how many bottles of ink does she require for one month?

A. 2

B. 5

C. 6

D. 12

E. 75

7. Use the diagram below to answer the question that follows.

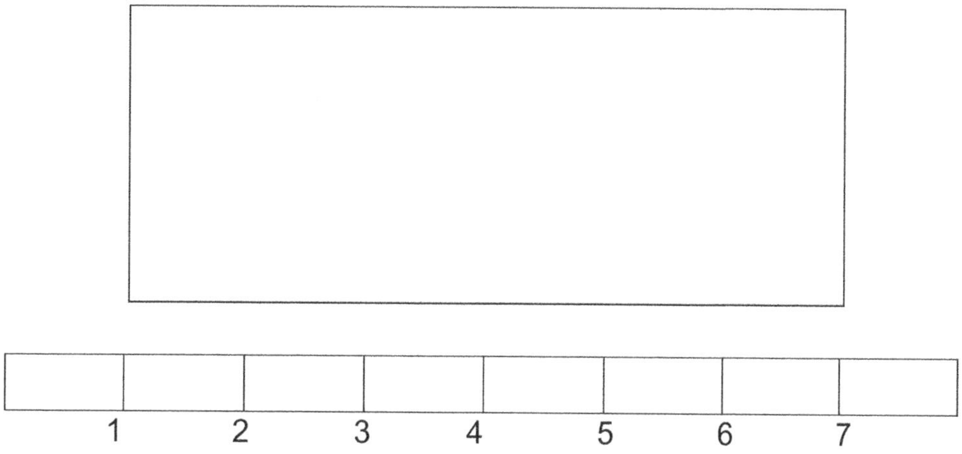

A scale drawing of a field is shown in the picture above. If the actual length of the field is 120 yards, what is the scale of the diagram?

A. 1 unit = 6 yards

B. 1 unit = 7 yards

C. 1 unit = 15 yards

D. 1 unit = 17 yards

E. 1 unit = 20 yards

8. A photograph measures 4 inches by 6 inches. Tom wants to make a wooden frame for the photo. He needs an extra inch of wood at each of the four corners in order to finish off the edges. What total length of wood will he need in order to complete the project?

A. 10 inches

B. 12 inches

C. 16 inches

D. 20 inches

E. 24 inches

9. It takes Martha 4 hours and 10 minutes to knit one woolen cap. At this rate, how long will it take her to knit 12 caps?

A. 40 hours

B. 42 hours

C. 46 hours

D. 48 hours

E. 50 hours

10. The Jones family needs to dig a new well. The well will be 525 feet deep, and it will be topped with a windmill which will be 95 feet in

height. What is the distance from the deepest point of the well to the top of the windmill?

A. 95 feet

B. 430 feet

C. 525 feet

D. 610 feet

E. 620 feet

11. Mr. Martin receives the following test report for a student who is in the tenth grade:

Raw Score	Percentile	Stanine	Grade Equivalent
68	54	6	9.2

Which of the following statements provides a correct interpretation of the student's results?

A. The student will be placed in the ninth grade.

B. 68% of the other students taking the test scored better than this student.

C. This student scored as well as or better than 32% of the other students taking the test.

D. This student scored as well as or better than 54% of the other students taking the test.

E. This student scored as well as or better than 46% of the other students taking the test.

12. At an elementary school, 3 out of ten students are taking an art class. If the school has 650 students in total, how many total students are taking an art class?

A. 65

B. 130

C. 195

D. 217

E. 325

13. Mrs. Emerson plays a card game with the children in her class. She has 12 cards that have a picture of a fish, 15 cards that have a picture of a dog, 25 cards that have a picture of a cat, and 18 cards that have picture of a rabbit. She draws cards from the deck at random and shows them to the class. If the first card she draws is a rabbit, what is the probability that the next card will be a cat or a rabbit?

A. $^{25}/_{69}$

B. $^{25}/_{70}$

C. $^{42}/_{69}$

D. $^{43}/_{69}$

E. $^{43}/_{70}$

14. What is the best estimate for $1{,}198 \div 29$?

A. 37

B. 40

C. 60

D. 400

E. 600

15. Simplify the following expression: $-243 - (+225) + 13$

A. 5

B. –5

C. –31

D. –455

E. –481

16. What is the remainder for the following: $251 \div 13$?

A. 3

B. 4

C. 5

D. 13

E. 19

17. Mrs. Thompson is having a birthday party for her son. She is going to give balloons to the children. She has one bag that contains 13 balloons, another that contains 22 balloons, and a third that contains 25 balloons. If 12 children are going to attend the party including her son, and the total amount of balloons is to be divided equally among all of the children, how many balloons will each child receive?

A. 3

B. 4

C. 5

D. 6

E. 7

18. Mount Pleasant is 15,238 feet high. Mount Glacier is 9,427 feet high. Which of the following is the best estimate of the difference between the altitudes of the two mountains?

A. 5,700

B. 5,800

C. 5,900

D. 6,000

E. 6,100

19. A bookstore is offering a 15% discount on books. Janet's purchase would be $90 at the normal price. How much will she pay after the discount?

A. $75.50

B. $76.50

C. $77.50

D. $85.50

E. $86.50

20. John is measuring plant growth as part of a botany experiment. Last week, his plant grew $7\frac{3}{4}$ inches, but this week his plant grew $10\frac{1}{2}$ inches. What is the difference in growth in inches between the two weeks?

A. $2\frac{1}{4}$ inches

B. $2\frac{1}{2}$ inches

C. $2\frac{3}{4}$ inches

D. $3\frac{1}{4}$ inches

E. $3\frac{1}{2}$ inches

21. At the beginning of a class, one-fourth of the students leave to attend band practice. Later, one half of the remaining students leave to go to PE. If there were 15 students remaining in the class at the end, how many students were in the class at the beginning?

A. 30

B. 40

C. 45

D. 50

E. 80

22. Patty works 23 hours a week at a part time job for which she receives $7.50 an hour. She then gets a raise, after which she earns $184 per week. She continues to work 23 hours per week. How much did her hourly pay increase?

A. 50 cents an hour

B. 75 cents an hour

C. $1.00 an hour

D. $8.00 an hour

E. $11.50 an hour

23. A packaging company places tape around a package that measures 4 inches in height, 5 inches in width, and 18 inches in length. One continuous piece of tape is placed around all four sides: on the top, bottom and both ends. Two further continuous pieces of tape are placed through the middle of the package around all four sides as shown in the illustration. How much tape is needed for this package?

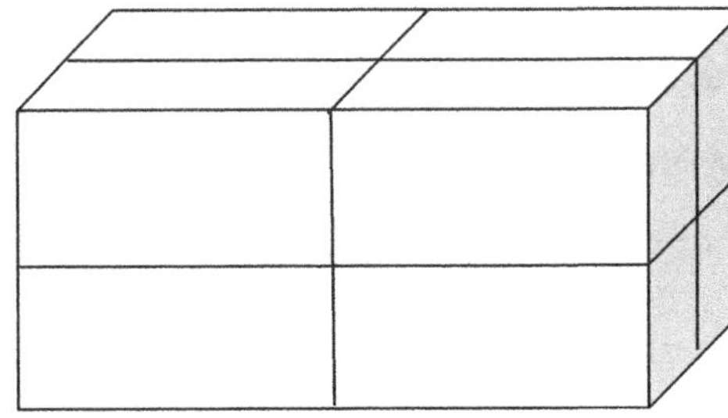

A. 16 inches

B. 20 inches

C. 72 inches

D. 100 inches

E. 108 inches

24. Sheng Li is driving at 70 miles per hour. At 10:00 am, he sees this sign:

Washington	**140 miles**
Yorkville	**105 miles**
Zorster	**210 miles**

He continues driving at the same speed. Where will Sheng Li be at 11:00 am?

A. 70 miles from Washington

B. 105 miles from Washington

C. 75 miles from Yorkville

D. 80 miles from Yorkville

E. 150 miles from Zorster

25. Mayumi spent the day counting cars for her job as a traffic controller. In the morning she counted 114 more cars than she did in the afternoon. If she counted 300 cars in total that day, how many cars did she count in the morning?

A. 90

B. 93

C. 114

D. 186

E. 207

26. Solve for x: $3x + 5 - 2x = 15$

A. 2

B. 3

C. 4

D. 5

E. 10

27. Shania is entering a talent competition which has three events. The third event (C) counts three times as much as the second event (B), and the second event counts twice as much as the first event (A). Which equation below can be used to calculate Shania's final score for the competition?

A. A + 2B + C

B. A + 2B + 3C

C. A + 3B + 2C

D. A + 2B + 6C

E. A + B + C

28. Tiffany buys five pairs of socks for \$2.50 each. The next day, she decides to exchange these five pairs of socks for four different pairs that cost \$3 each. She uses this equation to calculate her refund:

$(5 \times \$2.50) - (4 \times \$3)$

Which equation below could she have used instead?

A. $(5 \times 4) - (3 \times 2.50)$

B. $\$2.50 - 4(\$3 - \$2.50)$

C. $(5 \times 4) + (3 \times 2.50)$

D. $\$3 - (4 \times \$2.50)$

E. $\$3 - (5 \times \$2.50)$

29. Mr. Carlson needs to calculate 35% of 90.

To do so, he uses the following equation:

$$\frac{35 \times 90}{100}$$

Which of the following could he also have used?

A. $(35 \times 90) \div 100$

B. $(35 \div 90) \times 100$

C. $(35 - 90) \times 100$

D. $90 \times .0035$

E. $90 \div 35$

30. Read the information in the box below and answer the question that follows:

> - A health and beauty store has 90 bottles of shampoo for sale when the store opens for business on Monday morning.
> - These 90 bottles of shampoo consist of 15 bottles of strawberry-scented shampoo, 25 bottles of rose-scented shampoo, and 50 bottles of unscented shampoo.
> - At the close of business on Monday, 18 bottles of rose-scented shampoo remain in the store.

Which of the following facts can be determined from the information above?

A. The quantity of shampoo that the store normally offers for sale.

B. The average price of a bottle of shampoo.

C. The quantity of strawberry-scented shampoo sold on Monday.

D. The quantity of rose-scented shampoo sold on Monday.

E. The total quantity of shampoo left in the store at the close of business on Monday.

31. Which one of the following statements is correct?

A. $^5/_6 > {^5/_9} > {^2/_{10}}$

B. $^2/_6 > {^5/_8} > {^5/_6}$

C. $^2/_9 > {^5/_9} > {^2/_6}$

D. $^5/_9 > {^5/_6} > {^2/_9}$

E. $^2/_6 > {^2/_9} > {^5/_9}$

32. Use the mathematical expression below to answer the question that follows.

$^1/_6 < ? < {^4/_6}$

Which of the following fractions would correctly complete the expression above?

A. $^1/_3$

B. $^1/_9$

C. $^2/_3$

D. $^6/_9$

E. $^8/_{12}$

33. Read the problem below and then answer the question that follows.

Tom and Mary are planning a cross-country trip. They plan to drive 300 miles each day for seven days. Their car can travel 25 miles on one gallon of gasoline. How much money in total will they need to pay for gasoline during their trip?

What piece of information is needed in order to answer the problem?

A. The amount of gasoline that the tank of the car can hold.

B. The total amount of miles that they will drive that week.

C. The price per gallon of gasoline.

D. The day of the week that their journey will begin.

E. The average speed of the car in miles per hour.

34. Read the problem below and then answer the question that follows.

Paul leaves his house at 5:30 to go running. He runs 2 miles north through town, then continues 3 miles north out of town. He then runs south to his house along the same route. What is Paul's running pace?

What piece of information is needed in order to answer the problem?

A. The amount of steps that Paul makes.

B. The time that Paul returns home.

C. The length of Paul's stride.

D. The length of the return journey.

E. The total distance round-trip.

35. Use the chart below to answer the question that follows.

a	*b*
1.25	2.25
1.50	3.50
1.75	
2.00	6.00
2.25	7.25
2.50	8.50

The chart above demonstrates the relationship between variables *a* and *b*. What is the value of *b* that is missing from the chart?

A. 4.25

B. 4.50

C. 4.75

D. 5.00

E. 5.25

36. If the value of variable *x* is between 0.003 and 0.63, which one of the following could be variable *x*?

A. 0.0020

B. 0.0060

C. 0.6350

D. 0.7405

E. 0.0006

37. Which of the following mathematical expressions is equal to $(x \times y) \div z$?

A. $(x \div y) \times z$

B. $(x \times z) \div y$

C. $(x \div z) \times y$

D. $(y \div x) \div z$

E. $z \times (x \div y)$

38. Which one of the following numbers is between 4,587,213 and 4,732,841?

A. 4,496,215

B. 4,567,633

C. 4,579,554

D. 4,587,125

E. 4,723,524

39. Use the information below to answer the question that follows.

- The supermarket is 12 miles away from the gas station.
- Tom's house is 18 miles away from the gas station.

Based on the information given above, which one of the following statements is correct?

A. Tom's house is 6 miles from the supermarket.

B. Tom's house is 12 miles from the supermarket.

C. Tom's house is no more than 18 miles from the supermarket.

D. Tom's house is exactly 18 miles from the supermarket.

E. Tom's house is no more than 30 miles from the supermarket.

40. Carl swam three races this week. The time of his first race was 36.21 seconds. The time of the second race was 35.78 seconds. The time of his third race was 34.93 seconds. If each of these times is rounded to the nearest one-tenth of a second, what is the estimate of Carl's total time for all three of the races?

A. 106 seconds

B. 106.8 seconds

C. 106.9 seconds

D. 107 seconds

E. 107.1 seconds

41. Fatima drives 21 miles round trip every day between her home and her office. If her daily journey is rounded to the nearest 5 miles, which of the following is the best estimate of the total miles that Fatima drives in ten days?

A. 150 miles

B. 200 miles

C. 210 miles

D. 250 miles

E. 300 miles

42. Use the information below to answer the question that follows.

Classes will be held every Wednesday morning. If there are fewer than 3 children present for a class, the class will be canceled. If there is inclement weather, the class will be canceled.

It is Wednesday morning and the class has been canceled. Which one of the following statements is correct?

A. Fewer than three children were present for the class.

B. There was inclement weather.

C. Fewer than three children were present for the class and there was inclement weather.

D. Fewer than three children were present for the class or there was inclement weather.

E. More than three children were present for the class or there was inclement weather.

43. Use the information below to answer the question that follows.

> If the distance from his house to his destination is 5 miles or more, Jose uses his motorcycle.
> If the distance from his house to his destination is less than 5 miles but more than 1 mile, Jose uses his bicycle.
> If the distance from his house to his destination is 1 mile or less, Jose walks.

Jose uses his bicycle to go to Manuel's house. Which one of the following statements could be true?

A. Manuel's house is 1 mile or less from Jose's house.

B. Manuel and Jose live 8 miles apart.

C. Jose's house is at least 6 miles from Manuel's.

D. Jose lives 4 miles from Manuel.

E. The round trip between Jose and Manuel's houses is more than 10 miles.

44. Solve for x: $-12x + 15 + 16x = 31$

A. 4

B. 5

C. 6

D. 8

E. 15

45. Use the information below to answer the question that follows.

> Classes in the morning last for 45 minutes, but classes in the afternoon last for 50 minutes.
>
> Lunch begins promptly at 12:30 pm and finishes promptly at 1:00 pm.
>
> There are 3 classes after lunch and 4 classes before lunch.
>
> There are no breaks between classes or between classes and lunch.

Which one of the following statements could be true?

A. Classes begin at 9:30am.

B. Classes begin at 10:00am.

C. The second class after lunch begins at 2:00pm.

D. The second class after lunch begins at 2:50pm.

E. The third class after lunch begins at 3:00pm.

46. Use the table below to answer the question that follows.

Waterloo Station Bus Timetable	
Departure Time	Arrival Time
9:18 am	11:06 am
10:32 am	12:20 pm
11:52 am	?
1:03 pm	2:51 pm

The bus journeys from Waterloo Station to a nearby town are always the same duration. What time is missing from the above timetable?

A. 12:40 pm

B. 1:34 pm

C. 1:40 pm

D. 1:48 pm

E. 1:51 pm

47. Use the chart below to answer the question that follows.

Diesel Oil Consumption in New York City and all of New York State in Tons		
Year	**New York City**	**New York State**
1989	1528	6547
1994	1782	6118
1999	1693	5974
2004	1521	6128
2009	1844	7029
2014	1732	8192

What was the diesel oil consumption in tons for all areas in the state outside of New York City for 1999?

A. 1693

B. 5974

C. 4281

D. 5019

E. 6547

48. Use the graph below to answer the question that follows.

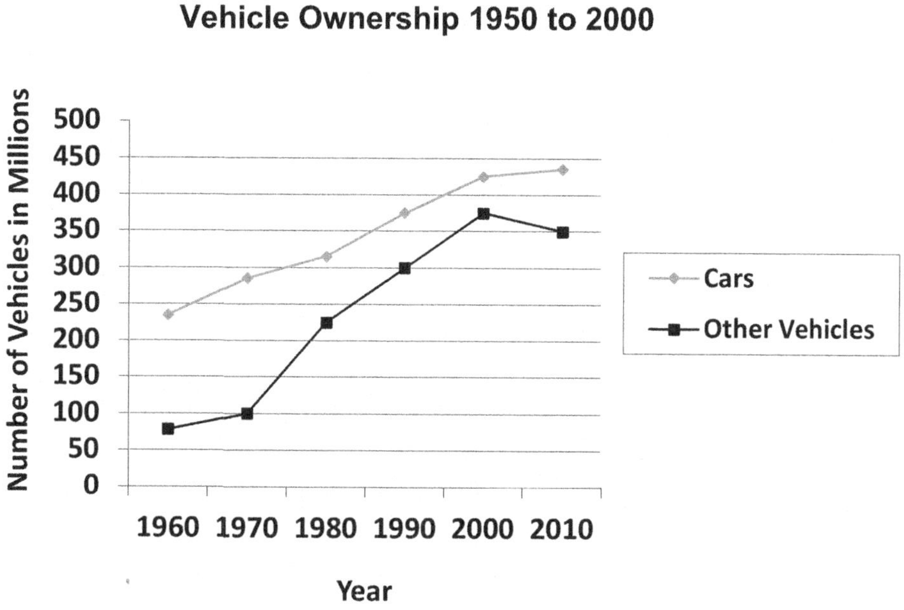

During what year was there the smallest difference between the numbers of cars owned and the number of other vehicles owned?

A. 1960

B. 1970

C. 1980

D. 1990

E. 2000

49. Use the chart below to answer the question that follows.

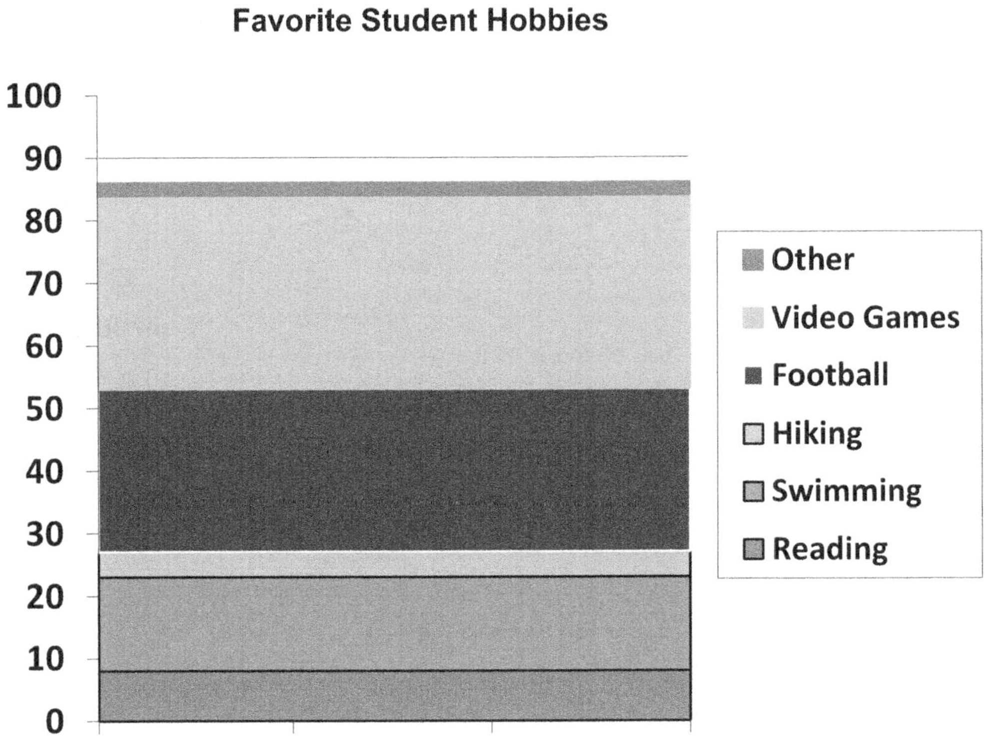

The percentage of students who enjoy participating in social media is omitted from the above chart. What percentage of students represented in the above chart have participating in social media as a favorite hobby?

A. 2%

B. 14%

C. 19%

D. 22%

E. 24%

50. Use the graph below to answer the question that follows.

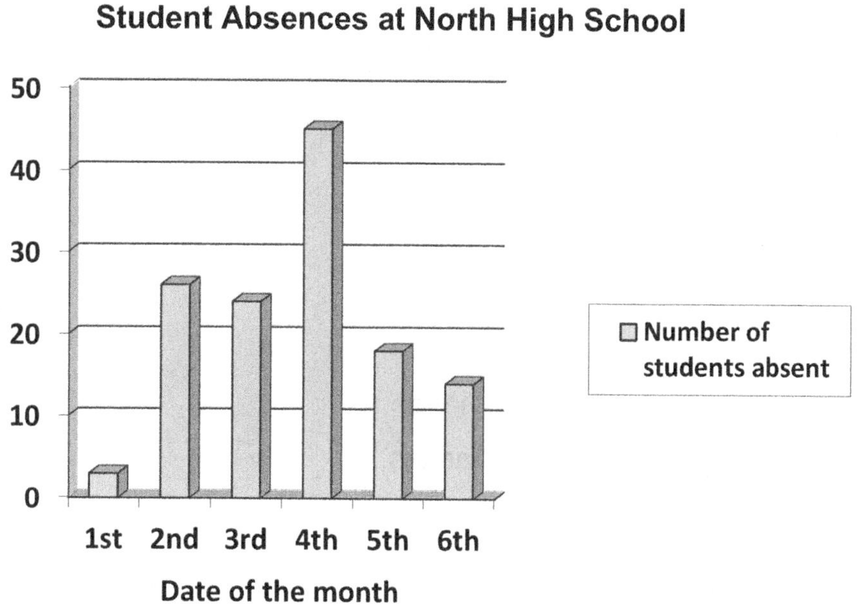

North High School experiences a flu epidemic. Between which dates does the number of student absences change the most?

A. the first to the second

B. the second to the third

C. the third to the fourth

D. the fourth to the fifth

E. the fifth to the sixth

CBEST Math Practice Test 3 – Answers

1. The correct answer is B.

He owns 26 yachts and needs 6 feet 10 inches of rope for each one.

Convert the feet and inches measurement to inches.

6 feet 10 inches =

$(6 \times 12) + 10$ inches =

$72 + 10 = 82$ inches

Then multiply by the number of items.

$26 \times 82 = 2132$ inches of rope needed

Then convert back to feet and inches.

2132 inches $\div$ 12 = 177 feet 8 inches

2. The correct answer is C.

Count how many blocks lie along the outer edges of the shaded area in order to get your result.

Top boundary = 4 feet

Left side boundary = 5 feet

Bottom boundary = 3 feet

Right boundary = 6 feet (Don't forget to count the piece shaped like the upside-down "L" on the right.)

Then add these amounts to get your result.

$4 + 5 + 3 + 6 = 18$ feet

3. The correct answer is C.

To get the average, add up all of the items.

$98 + 78 + 89 + 85 + 90 = 440$

There are five scores, so there were five tests.

Divide the total points by the number of tests in order to get the average.

440 ÷ 5 = 88

4. The correct answer is A.

Sugar is a dry item, so it is measured by weight. Remember that wet items are measured in pints and quarts, while dry items are measured in ounces and pounds, or tons in the case of extremely large quantities.

Feet and inches are linear measurements; they are not used for weight.

A bag is sugar is a small item, so the correct answer is ounces.

5. The correct answer is B.

First of all, add up the number of questions answered correctly.

12 + 20 + 32 + 32 = 96

Then add up the total number of questions.

15 + 25 + 35 + 45 = 120

Now divide the number of questions answered correctly by the total number of questions to get her percentage score.

96 ÷ 120 = 80%

6. The correct answer is D.

Assuming there are 30 days in the month, we can divide as shown.

30 days ÷ 5 days per 2 bottles =

6 days × 2 bottles = 12 bottles needed for 1 month

7. The correct answer is E.

STEP 1: Determine the length of the field in units.

We can see that the right-hand side of the field is on the number 7.

However, the field is not positioned over the first unit of the ruler.

So, the field is 6 units long.

STEP 2: Divide the actual length of the field by the units.

120 ÷ 6 = 20

STEP 3: Express the answer in units and yards.

1 unit = 20 yards

8. The correct answer is E.

Measure the length along the top and bottom of the frame, as well as the length of both sides in order to get the basic perimeter.

Top = 4 inches

Bottom = 4 inches

Left side = 6 inches

Right side = 6 inches

Total perimeter: 4 + 4 + 6 + 6 = 20 inches

Now add in the 4 extra inches for the four corners.

20 + 4 = 24 inches

9. The correct answer is E.

STEP 1: Convert into minutes the amount of time required to make one cap.

4 hours and 10 minutes =

(4 × 60) + 10 =

240 + 10 = 250 minutes needed to make one cap

STEP 2: Multiply by the total output.

250 minutes × 12 caps = 3000 minutes

STEP 3: Convert the total amount of minutes back to hours and minutes

3000 minutes ÷ 60 = 50 hours

10. The correct answer is E.

Add the feet above ground to the feet below ground to get the total distance.

525 + 95 = 620 feet

11. The correct answer is D.

Remember the following basic aspects of standardized score reports.

Raw score is the number of questions answered correctly, assuming that each correct response receives one point.

Percentile means that the student scored as well as or better than this percentage of other students taking the test.

Stanine is short for "Standard nine." Stanine shows the student's score on a scale of 1 to 9.

The grade equivalent shows the grade-level of school the student's performance. For example, a grade equivalent of 9.2 means that the student performed on the exam at the level of a 9th grade student.

In our case, we cannot say with certainty that this student will be placed in ninth grade classes since this depends on the school's policies.

The report states that the student's percentile score is 54. Therefore, the student scored as well as or better than 54% of the other students taking the test.

12. The correct answer is C.

Three out of ten students are taking the class. So, here we have the proportion 3 to 10.

STEP 1: Divide the total number of students by the second number in the proportion to get the number of groups.

650 ÷ 10 = 65 groups

STEP 2: Multiply the number of groups by the first number in the proportion in order to get the result.

3 × 65 = 195 art students

13. The correct answer is C.

Here is another probability question.

STEP 1: Remember that your first step is to determine the number of items in the data set, before any items are removed.

Mrs. Emerson has 12 cards that have a picture of a fish, 15 cards that have a picture of a dog, 25 cards that have a picture of a cat, and 18 cards that have picture of a rabbit.

12 + 15 + 25 + 18 = 70

STEP 2: Determine the amount in the data set after any items have been removed.

We know that the first card she draws is a rabbit, so she has taken one item from the data set.

70 – 1 = 69

STEP 3: Determine the amount in the subset.

Our subset is cards with cats or rabbits. Before any items were removed, we had 25 cards with picture of a cat, and 18 cards with picture of a rabbit. Then one card with a rabbit was removed. So, add and subtract as shown.

25 + 18 – 1 = 42 cat or rabbit cards remaining

STEP 4: The probability is expressed as a fraction, with the subset on the top and the data set on the bottom.

$^{42}/_{69}$

14. The correct answer is B.

We can see from the answer choices that we are rounding to the nearest increment of 10.

1,198 is rounded up to 1,200

29 is rounded up to 30

Now do the operation.

1,200 ÷ 30 = 40

15. The correct answer is D.

Remember to be careful with the negatives.

–243 – (+ 225) + 13 =

–243 – 225 + 13 =

–468 + 13 =

–455

16. The correct answer is B.

251 ÷ 13 = ?

Do long division as shown in order to get the remainder.

$$\begin{array}{r} 19 \\ 13\overline{)251} \\ \underline{13} \\ 121 \\ \underline{117} \\ 4 \end{array}$$

17. The correct answer is C.

STEP 1: Add the items together to get the total amount of items available.

13 + 22 + 25 = 60 balloons in total

STEP 2: Divide the amount of items available by the number of people.

60 ÷ 12 = 5

18. The correct answer is B.

We have to subtract to find the difference in height between the two mountains.

First of all, round each number up or down.

Looking at the answer choices, we can see that we need to round to the nearest increment of 100.

15,238 is rounded down to 15,200

9,427 is rounded down to 9,400

Now subtract to get your answer.

15,200 – 9,400 = 5,800

19. The correct answer is B.

STEP 1: Determine the value of the discount by multiplying the normal price by the percentage discount.

$90 × 15% = $13.50 discount

STEP 2: Subtract the value of the discount from the normal price to get the new price.

$90 – $13.50 = $76.50

20. The correct answer is C.

STEP 1: You can express the fractions as decimals for the sake of simplicity.

$10\frac{1}{2} = 10.50$

$7\frac{3}{4} = 7.75$

STEP 2: Then subtract to find the increase.

10.50 – 7.75 = 2.75

STEP 3: Then convert back to a mixed number.

$2.75 = 2\frac{3}{4}$

21. The correct answer is B.

One-fourth of the students leave to attend band practice. Later, one half of the remaining students leave to go to PE.

We know we have 15 students remaining in class after the others have left.

So, work backwards based on the facts given.

STEP 1: We have 15 students just after half of the students have left to go to PE, so divide 15 by one-half. So, if x represents the number of remaining students after the group of students have left for PE, we have the following equation.

$15 = x - (x \times \frac{1}{2})$

$15 = x - \frac{1}{2}x$

$15 = \frac{1}{2}x$

$15 \times 2 = \frac{1}{2}x \times 2$

$30 = x$

So, there were 30 students in class before PE class

STEP 2: We have 30 students after ¼ of them have gone to band, so we have the following equation.

$30 = x - (x \times \frac{1}{4})$

$30 = x - \frac{1}{4}x$

$30 = \frac{3}{4}x$

$30 \times 4 = \frac{3}{4}x \times 4$

$120 = 3x$

$120 \div 3 = 3x \div 3$

$40 = x$

So, there were 40 students in the class at the beginning.

22. The correct answer is A.

After her raise, she earns $184 per week. She continues to work 23 hours per week.

STEP 1: Determine the new hourly rate.

$184 ÷ 23 hours = $8 per hour

STEP 2: Determine the change in the hourly rate.

$8 - $7.50 = 50 cents per hour

23. The correct answer is E.

STEP 1: Label the dimensions on the diagram.

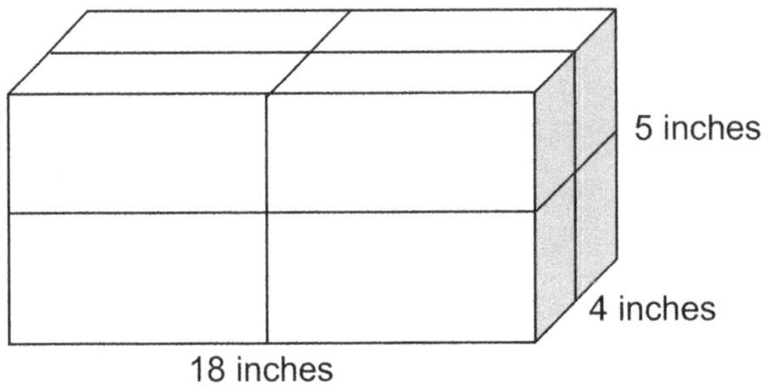

STEP 2: We are covering all four sides, so we need to multiply each of the dimensions by 4.

18 × 4 = 72

5 × 4 = 20

4 × 4 = 16

STEP 3: Add all of the above figures together to get your result.

72 + 20 + 16 = 108

24. The correct answer is A.

STEP 1: Determine the distance traveled.

If he is traveling 70 miles an hour, he will have traveled 70 miles after one hour has passed.

STEP 2: Determine the distance from the towns listed on the sign, considering that he has traveled for one hour.

Washington: 140 – 70 = 70 miles from Washington

Yorkville: 105 – 70 = 35 miles from Yorkville

Zorster: 210 – 70 = 140 miles from Zorster

STEP 3: Compare the above figures to your answer choices to get your result. After an hour, he is 70 miles from Washington, so A is correct.

25. The correct answer is E.

STEP 1: Subtract the excess from the total.

$300 - 114 = 186$

STEP 2: Allocate the difference into its respective parts.

We are dividing the day into two parts: morning and afternoon. There were 186 cars in total without the excess, so divide this into two parts.

$186 \div 2 = 93$

STEP 3: Determine the amount for the larger part.

There were 114 more cars in the morning, so add this back in.

$93 + 114 = 207$ cars in the morning

26. The correct answer is E.

Remember to deal with the integers, and then deal with the variable.

$3x + 5 - 2x = 15$

$3x + 5 - 5 - 2x = 15 - 5$

$3x - 2x = 10$

$x = 10$

27. The correct answer is D.

Work out the equation based on the facts provided in the problem.

The second event (B) counts twice as much as the first event (A), so we need to represent the value of the second event as 2B.

The third event (C) counts three times as much as the second event, so we need to multiply the value of the second event by 3.

$2 \times 3 = 6$

So, the value of the third event is 6C.

Therefore, the equation is A + 2B + 6C.

28. The correct answer is B.

Here is another exchange problem.

STEP 1: Think about the value of the four pairs of socks she is getting in the exchange. These socks cost 50 cents more each than the pairs she has already bought. So, we can express the difference in value of those four pairs of socks as: 4 × ($3 - $2.50)

STEP 2: Take into account the value of the extra pair of socks. She paid $2.50 for a fifth pair of socks, but she is only getting four pairs back on the exchange, so she is owed money back for that part of the purchase.

Therefore, we can calculate the refund owing as $2.50 – 4($3 - $2.50)

29. The correct answer is A.

The line in any fraction can be treated as the division symbol. Accordingly, we can divide by the denominator, which is 100 in this case.

$$\frac{35 \times 90}{100} = (35 \times 90) \div 100$$

30. The correct answer is D.

We don't know how many bottles of strawberry or unscented shampoo were sold. Nor do we know what the store sells normally. So, we cannot calculate the total quantity of shampoo left unsold in the store when it closes on Monday. We can only calculate the quantity of rose-scented shampoo sold since the facts tell us how many bottles of rose-scented shampoo remain in the store at the close of business.

31. The correct answer is A.

Rule each answer choice out by the process of elimination.

Remember these shortcuts when dealing with fractions:

Fractions with the same numerators:

If the numerators on the tops of two fractions are the same, the fraction with the smaller denominator is actually the greater fraction.

For example: $\frac{1}{2} > \frac{1}{4}$

If any of the inequalities in the answer choices have the same numbers in their numerators, you can then just compare the denominators in the answer choices in order to determine which fraction is greater.

This is the case with answer choices B and D, so let's evaluate them.

$^2/_6 > {}^5/_8 > {}^5/_6$ – Answer B is incorrect because $^5/_8$ is less than $^5/_6$.

$^5/_9 > {}^5/_6 > {}^2/_9$ – Answer D is incorrect because $^5/_9$ is less than $^5/_6$.

Fractions with the same denominators:

On the other hand, if the denominators on the bottoms of two fractions are the same, the fraction with the larger numerator is the greater fraction.

This is the case with answer choices C and E, so let's look at them next.

$^2/_9 > {}^5/_9 > {}^2/_6$ – Answer C is incorrect because $^2/_9$ is less than $^5/_9$.

$^2/_6 > {}^2/_9 > {}^5/_9$ – Answer E is also incorrect for the same reason.

Now have a look at the other answer choices.

Answer choice A is as follows: $^5/_6 > {}^5/_9 > {}^2/_{10}$

Using the principles above, we can see that $^5/_6 > {}^5/_9$ from answer A is correct.

So, next you can evaluate whether $^5/_9$ > $^2/_{10}$ from answer A is also correct.

First, we have to find the lowest common denominator.

$$\frac{5}{9} > \frac{2}{10}$$

$$\left(\frac{5}{9} \times \frac{10}{10}\right) > \left(\frac{2}{10} \times \frac{9}{9}\right)$$

$$\frac{50}{90} > \frac{18}{90}$$

Both parts of the inequality are correct, so A is the correct answer.

32. The correct answer is A.

Check each answer option one by one.

From answer A, we can see that $^1/_6$ is less than $^1/_3$.

Then put the other part of the inequality into the LCD to check the answer.

$$\frac{1}{3} < \frac{4}{6}$$

$$\left(\frac{1}{3} \times \frac{2}{2}\right) < \frac{4}{6}$$

$$\frac{2}{6} < \frac{4}{6}$$

Both parts of the inequality are correct, so answer A is the correct answer.

33. The correct answer is C.

In order to solve this problem, we would need to multiply the number of gallons of gasoline used per day by the cost of gasoline per gallon by the number of days traveled in order to calculate the total cost.

From these required facts, we are lacking the price of gasoline per gallon.

34. The correct answer is B.

We know that Paul will have run ten miles when he finishes since he runs 5 miles north, then returns and goes 5 miles south.

The question is asking about his running pace or speed.

In order to know speed, we need to know the distance traveled and the amount of time it takes to travel the distance.

So, we know the distance, but not the time.

Accordingly, we would need to know what time he gets back home in order to solve the problem.

35. The correct answer is C.

Remember to check the relationship between the numbers in each column on a row-by-row basis.

Here, we can solve by addition. The pattern is that the row number is added to the value in column *a* in order to find its value in column *b*.

So, for row 1: $1.25 + 1 = 2.25$

For row 2: $1.50 + 2 = 3.50$

For row 3: $1.75 + 3 = 4.75$

36. The correct answer is B.

Remember to line all of the numbers up in a column if you have difficulties solving problems like this visually.

By visual inspection, we can see that answers A and E are too small, while answers C and D are too large.

$0.003 < 0.006 < 0.63$, so answer B is correct.

37. The correct answer is C.

Remember that the division symbol is the same calculation mathematically as the line in a fraction.

So, express the equations as fractions to check the answer.

$$(x \times y) \div z = \frac{x \times y}{z} = \frac{xy}{z}$$

From answer C, we can determine that:

$$(x \div z) \times y = \frac{x}{z} \times y = \frac{xy}{z}$$

38. The correct answer is E.

Answers A to D are too small. Putting the numbers in a column, we can check that answer E is correct:

4,587,213

4,723,524

4,732,841

39. The correct answer is E.

For questions like this, you will recall that the points could lie on one continuous strait path like a line. Alternatively, the points could be laid out more like a triangle. However, the distance between points will always be greater when the points are linear.

If the points are linear, then the maximum distance will be calculated as follows:

12 miles + 18 miles = 30 miles

40. The correct answer is C.

Remember that the tenth is the decimal just to the right of the decimal point.

So, we need to round as required.

The first race was 36.21 seconds, which is rounded down to 36.2

The second race was 35.78 seconds, which is rounded up to 35.8

The third race was 34.93 seconds, which is rounded down to 34.9

Now add these figures together.

$36.2 + 35.8 + 34.9 = 106.9$

41. The correct answer is B.

We round the daily distance to 20, and then multiply by 10 to get the estimate of 200.

42. The correct answer is D.

The second fact tells us that if there are fewer than 3 children present for a class, the class will be canceled.

The third fact tells us that if there is inclement weather, the class will also be canceled.

43. The correct answer is D.

The second fact tells us that if the distance from his house to his destination is less than 5 miles but more than 1 mile, Jose uses his bicycle. If Jose uses his bicycle to go to Manuel's house, then it might be possible that Jose lives 4 miles from Manuel.

44. The correct answer is A.

$-12x + 15 + 16x = 31$

$-12x + 15 - 15 + 16x = 31 - 15$

$-12x + 16x = 16$

$4x = 16$

$4x \div 4 = 16 \div 4$

$x = 4$

45. The correct answer is A.

If classes last for 45 minutes and there are 4 classes before lunch, the morning classes last for 3 hours in total.

If lunch is at 12:30, it is therefore possible for classes to begin at 9:30.

46. The correct answer is C.

Each journey is 108 minutes (1 hour and 48 minutes) in duration.

So, we need to add 108 minutes to the departure time of 11:52 to get the arrival time of 1:40.

47. The correct answer is C.

Subtract the New York City amount from the total for the entire state for 1999 to get the answer.

5974 – 1693 = 4281

48. The correct answer is E.

Look to see which year has the smallest gap between the two lines.

49. The correct answer is B.

If the entire graph were complete, we would have 100%.

At the moment, we have 86%, so we need to subtract.

100% – 86% = 14%

50. The correct answer is D.

Notice that the question in asking about the biggest change in general, rather than the largest increase or the largest decrease. Visually, we can see that the gap between months 4 and 5 is the greatest.

CBEST Practice Math Test 4

1. Which of the following shows the numbers ordered from least to greatest?
   A. $-^1/_4$, $^1/_8$, $^1/_6$, 1
   B. $-^1/_4$, $^1/_8$, 1 , $^1/_6$
   C. $-^1/_4$, $^1/_6$, $^1/_8$, 1
   D. $-^1/_4$, 1 , $^1/_8$, $^1/_6$
   E. 1 , $^1/_6$, $^1/_8$, $-^1/_4$

2. When 1523.48 is divided by 100, which digit of the resulting number is in the tenths place?
   A. 1
   B. 2
   C. 3
   D. 4
   E. 5

3. You work as a climatologist and need to calculate the average high temperature in one city over a five-day period in degrees Celsius. However, the high temperatures are reported in Fahrenheit. You have collected the following data: Day 1: 72° F; Day 2: 68° F; Day 3: 65° F; Day 4: 82° F; Day 5: 81° F. What was the approximate average high temperature in degrees Celsius? °C = 0.56(°F – 32)
   A. 74°C
   B. 73°C
   C. 41°C
   D. 32°C
   E. 23°C

4. An art and craft store received $7,375 for sales of a certain type of scrapbook this year. If these scrapbooks were sold for $59 each, how many of them were sold this year?
   A. 135
   B. 125
   C. 120
   D. 75
   E. 65

5. One private airplane flew at a constant speed, traveling 780 miles in 2 hours. How many miles did this plane travel in the last 40 minutes of its journey?
   A. 120
   B. 180
   C. 200
   D. 260
   E. 240

6. A horse ran 12 furlongs in 2 minutes and 48 seconds. Assuming that the same amount of time was spent on each furlong, how many seconds does it take the horse to run one furlong?
   A. 0.014 seconds
   B. 0.14 seconds
   C. 1.40 seconds
   D. 14 seconds
   E. 140 seconds

7. A report shows that 2 out of every 20 employees of a particular company are interested in applying for a promotion. If the company has 480 employees in total, how many employees are interested in applying for a promotion?

A. 20
B. 24
C. 42
D. 48
E. 84

8. An item costs $22 each if the customer collects it in person from the store, and an extra $3 for postage and handling is charged per item if the customer wants the item sent by courier. This week, 32 customers purchased this item and requested that the item be sent by courier. How much money in total did the store collect on the items sold to these 32 customers?
A. $800
B. $704
C. $575
D. $525
E. $96

9. $^{6}/_{25}$ of the inventory has been sold this month. Approximately what percentage of the inventory has been sold?
A. 0.24%
B. 2.40%
C. 24.0%
D. 4.167%
E. 6.667%

10. A vat contains 163.75 units of red colorant, 107.50 units of blue colorant, 91.25 units of yellow colorant, and 10.30 units of black colorant. Which of the following represents, in terms of units, how full the vat is after these 4 colorants have been placed in it?
A. 362.50
B. 371.50
C. 372.80

D. 373.50
E. 374.50

11. A customer who owns a small hotel has ordered 10 new quilts. Each quilt requires 2 yards of red fabric for the front, 1 yard of blue fabric for the front, and a further 3 yards of blue fabric for the back. The quilts need to have an embellishment in gold, and a total amount of 6 yards of gold fabric is needed to make the embellishments for all 5 quilts. Each quilt also has edging in white, and half a yard of white material is needed for the edging for each quilt. How many yards of fabric in total will be needed to complete this order?
A. 7.7
B. 77
C. 3.85
D. 38.50
E. 3.95

12. A store sells domestic cleaning products. A certain type of liquid cleaner is sold in increments of 1/4 of a cup. Each 1/4 of a cup costs 50 cents. One customer buys $10^1/_4$ cups of this cleaner. How much will she pay for this purchase?
A. $5.13
B. $5.50
C. $10.50
D. $20.50
E. 21.50

13. The cost of sales figures each month for a company's first five months of business this year were: $723, $618, $576, $812, and $984. What is the best estimate of the total cost of sales for the first five months of business this year?

A. $2,600
B. $2.700
C. $3,600
D. $3,700
E. $4,700

14. A liquid ingredient is stored in 5-quart containers. There are two partially-full containers, one with $4^3/_8$ quarts and another with $3^7/_8$ quarts. How many quarts are there in total in these two containers?
A. $1^1/_4$
B. 7
C. $7^1/_8$
D. $8^1/_4$
E. 9

15. A small factory uses tarpaulin to make covers for farm implements. There was $12^7/_{16}$ yards of tarpaulin at the start of the day. At the end of the day, $8^9/_{16}$ yards of tarpaulin is left. Which amount below represents the amount of tarpaulin used this day in yards?
A. $2^{14}/_{16}$
B. $3^1/_8$
C. $3^7/_8$
D. $4^7/_8$
E. 5

16. Abdul purchased 80 items for sale, and he has sold 0.75 of them in relation to the total purchased. How many items does he have left after making these sales?
A. 10 items
B. 20 items
C. 25 items

D. 40 items
E. 42 items

17. A footwear store can purchase 325 pairs of tennis shoes from its normal supplier for $4 a pair. It can get the same 325 pairs of shoes from a second supplier for $1,250 plus 6% sales tax, or from a third supplier for $1,290. How much will the store pay to get the best deal?
A. $1,250.00
B. $1,290.00
C. $1,300.00
D. $1,367.40
E. $1,368.40

18. 4 out of every 5 employee-satisfaction questionnaires have been completed and returned. If a company has 250 total employees, and every employee must complete and return the questionnaire, how many questionnaires have not been completed and returned?
A. 4
B. 5
C. 50
D. 200
E. 210

19. A flower store sells poinsettia plants for $20 during December and for $12 during January. In December, 55 customers purchased poinsettias, and 20 customers purchased them in January. How much money did the store receive for poinsettia sales during December and January?
A. $240
B. $1,060
C. $1,100

D. $1,340
E. $1,360

20. In the last step of doing a calculation, Wei Li added 92 instead of subtracting 92. What shortcut can Wei Li perform in order to get the correct calculation?
A. Subtract 46 from his erroneous result.
B. Add 92 to his erroneous result.
C. Subtract 92 from his erroneous result.
D. Add 184 to his erroneous result.
E. Subtract 184 from his erroneous result.

21. A caterpillar travels 10.5 inches in 45 seconds. How far will it travel in 6 minutes?
A. 45 inches
B. 63 inches
C. 64 inches
D. 84 inches
E. 90 inches

22. The ratio of males to females in the senior year class of Carson Heights High School was 6 to 7. If the total number of students in the class is 117, how many males are in the class?
A. 48
B. 54
C. 56
D. 58
E. 63

23. Results from a questionnaire administered to customers of a particular supermarket show that 4 out of 7 customers prefer toffee-flavored ice cream to coffee-flavored ice cream. Based on these results, if 1,540 customers purchased one of these two flavors of ice cream, how many of them would have purchased coffee-flavored ice cream?

A. 220
B. 420
C. 560
D. 660
E. 880

24. An employment agency for temporary employees charges clients $15 per hour for each hour the temporary employee works. The agency pays each temporary employee $12 an hour and retains the difference as a commission. The agency had 10 employees who worked 40 hours each this week. How much did the agency make on commission for these 10 employees this week?

A. $30.00
B. $120.00
C. $1,200.00
D. $4,800.00
E. $5,200.00

25. 49 out of the 50 items in a company's product line had above average sales this month. What percentage of the items in the product line had above average sales this month?

A. 0.098%
B. 0.98%
C. 9.80%
D. 98%
E. 980%

26. Sales each day for the past five days have been as follows: $90, $85, $85, $105, $110. What was the daily average sales amount during this five-day period?

A. $25
B. $85
C. $90
D. $95
E. $105

27. A fabric store sells ribbon in 3-inch or one-foot increments. One customer wanted two types of ribbon, and she bought $8^3/_4$ feet of one type of ribbon and $7^1/_2$ feet of another type. How much ribbon did this customer buy in total?

A. 7 feet and 6 inches
B. 8 feet and 9 inches
C. 15 feet and 3 inches
D. 16 feet and 3 inches
E. 17 feet and 3 inches

28. Hours spent on a work order are recorded by the tenth of an hour in 6-minute increments. For a particular work order, $28^3/_{10}$ hours in total have been budgeted. $7^9/_{10}$ hours have already been spent on the work order. Which amount below represents the amount of time left for this work order?

A. $36^1/_5$
B. $35^6/_{10}$
C. $20^2/_5$
D. $21^3/_5$
E. $22^3/_5$

29. A measurement of 116 feet is how many inches longer than a measurement of 36 yards?

A. 8
B. 80
C. 96
D. 960
E. 3,744

30. The ages of 5 siblings are: 2, 5, 7, 12, and x. If the average age of the 5 siblings is 8 years old, what is the age (x) of the 5th sibling?

A. 8
B. 10
C. 12
D. 14
E. 16

31. A recipe of the ingredients needed to make 4 brownies is provided below.

Brownie recipe
¼ cup of flour
½ cup of sugar
¼ cup of butter
3 tablespoons of cocoa powder
¼ teaspoon of baking powder
½ teaspoon of vanilla extract

How much cocoa powder and baking powder together is needed to make 12 brownies? (1 tablespoon = 3 teaspoons)

A. 9¼ teaspoons
B. 27¼ teaspoons
C. 27½ teaspoons

D. $27\frac{3}{4}$ teaspoons
E. $28\frac{3}{4}$ teaspoons

32. Look at the table below and then answer the question that follows.

	Live in the City	Live in the Country
Factory workers	35	23
Office workers	42	20

The table above shows data on the distribution of residents of a particular county, according to their types of work and the locations of their residences. If one of the residents from this sample is selected at random, what is the probability that the resident will be an office worker who lives in the country?

A. $^{1}/_{7}$
B. $^{1}/_{6}$
C. $^{1}/_{3}$
D. $^{23}/_{120}$
E. $^{42}/_{120}$

33. An illusionist pulls colored scarves out of a hat at random. The hat contains 5 red scarves and 6 blue scarves. The other scarves in the hat are green. If a scarf is pulled out of the hat at random, the probability that the scarf is red is $^{1}/_{3}$. How many green scarves are in the hat?

A. 3
B. 4
C. 5
D. 6
E. 7

34. $\frac{3}{4}x - 2 = 4$, x = ?

A. $\frac{8}{3}$

B. $\frac{1}{8}$

C. 8

D. –8

E. 24

35. Which of the following steps will solve the equation for x: $4x - 3 = 2$

A. Add 3 to both sides of the equation, and then divide both sides by 4.

B. Add 3 to both sides of the equation, and then subtract 4 from both sides.

C. Add 2 to both sides of the equation, and then divide both sides by 4.

D. Subtract 2 from both sides of the equation, and then divide both sides by –3.

E. Divide both sides of the equations by 4, and then subtract 3 from both sides.

36. Toby is going to buy a car. The total purchase price of the car is represented by variable C. He will pay D dollars immediately, and then he will make equal payments (P) each month for a certain number of months. Which equation below represents the amount of his monthly payment (P)?

A. $\frac{C-D}{M}$

B. $\frac{C}{M} - D$

C. $\frac{M}{C-D}$

D. $D - \frac{C}{M}$

E. $\frac{C}{M}$

37. A baseball team sells T-shirts and sweatpants to the public for a fundraising event. The total amount of money the team earned from these sales was $850. Variable t represents the number of T-shirts sold and variable s represents the number of sweatpants sold. The total sales in dollars is represented by the equation $25t + 30s$. The amount earned by selling sweatpants is what fraction of the total amount earned?
    A. $s/850$
    B. $30s/850$
    C. $(25t + 30s)/850$
    D. $t/850$
    E. $25t/850$

38. The total funds, represented by variable F, available for P charity projects is represented by the equation F = $500P + $3,700. If the charity has $40,000 available for projects, what is the greatest number of projects that can be completed?
    A. 72
    B. 73
    C. 74
    D. 79
    E. 80

39. If $3x - 9 = -18$, then $x = ?$

A. –6
B. 6
C. –3
D. 3
E. 2

40. Read the information below and answer the question that follows.

- If door A is locked with the red key, then door B is locked with the blue key.
- If door B is locked with the blue key, then door C is locked with the green key.
- The key that locks door B also locks door D.

If door A is locked with the red key, then which of the following must be true?

A. Door C is locked with the blue key.
B. Door C is locked with the red key.
C. Door D is locked with the blue key.
D. Door D is locked with the red key.
E. Door B is currently unlocked.

41. The area of a square floor is 64 square units. The floor needs to be covered entirely with tiles. Each floor tile is 4 square units. How many tiles are needed to cover the floor?

A. 8
B. 12
C. 16
D. 24
E. 36

42. Which of the following dimensions would be needed in order to find the area of the figure?

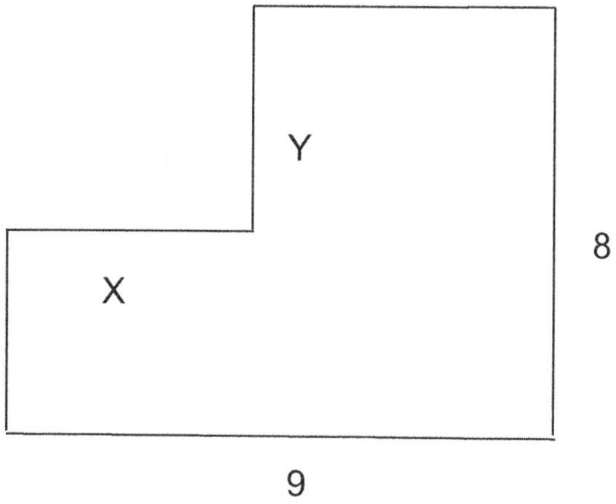

A. X only
B. Y only
C. Both X and Y
D. Either X or Y
E. Neither X nor Y

Go on to the next page.

43. Brooke wants to put new flooring in her living room. She will buy the flooring in square pieces that measure 1 square foot each. The entire room is 12 feet by 16 feet. The bookcases are one and a half feet deep from front to back. Flooring will not be put under the bookcases. Each piece of flooring costs $4.25. A diagram of her living room is provided.

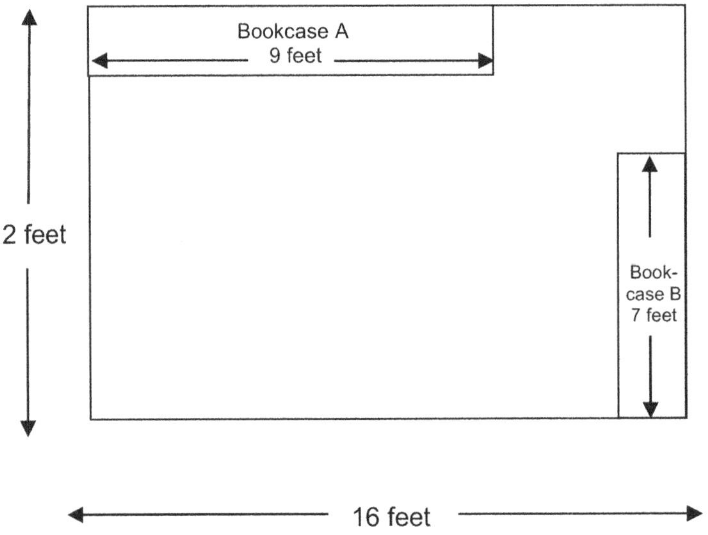

How much will Brooke pay to cover her living room floor? Provide the cost to the nearest dollar.

A. 168
B. 192
C. 714
D. 758
E. 816

44. A football field is 100 yards long and 30 yards wide. What is the area of the football field in square yards?

A. 3000
B. 1500

C. 300
D. 260
E. 240

45. A small pasture has a length of 5 yards and a width of 3 yards. Barbed wire will be placed on all four sides of the outside of this pasture. How many yards of barbed wire should be ordered?
A. 15
B. 16
C. 18
D. 40
E. 42

46. The graph below shows the relationship between the total number of hamburgers a restaurant sells and the total sales in dollars for the hamburgers. What is the sales price per hamburger?

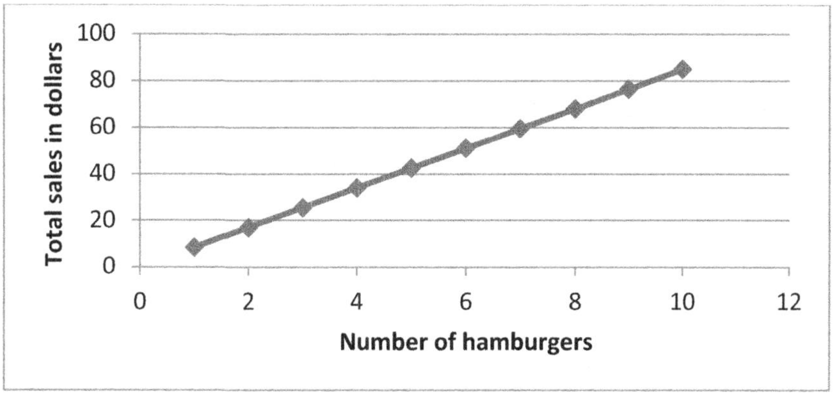

A. $4.00
B. $8.00
C. $8.50
D. $9.50
E. $10.00

47. In Brown County Elementary School, parents are advised to have their children vaccinated against five childhood diseases. According to the chart below, how many children were vaccinated against at least three diseases?

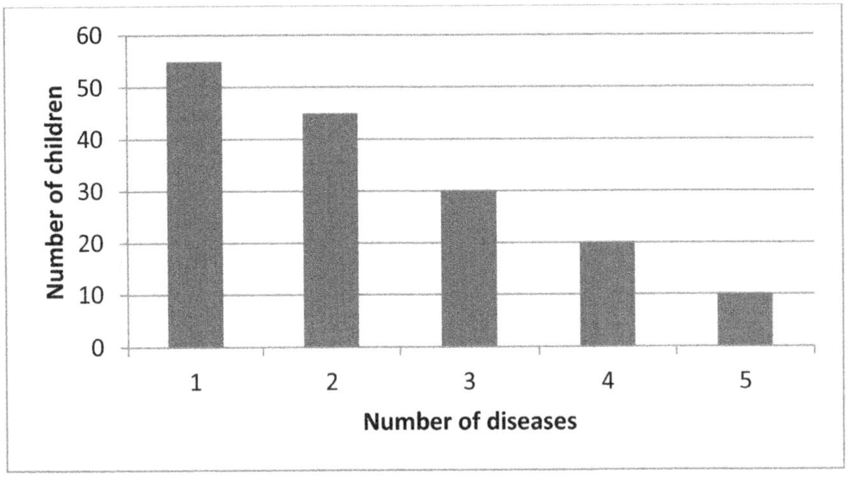

A. 30
B. 50
C. 60
D. 100
E. 130

48. A zoo has reptiles, birds, quadrupeds, and fish. At the start of the year, they have a total of 1,500 creatures living in the zoo. The pie chart below shows percentages by category for the 1,500 creatures at the start of the year.

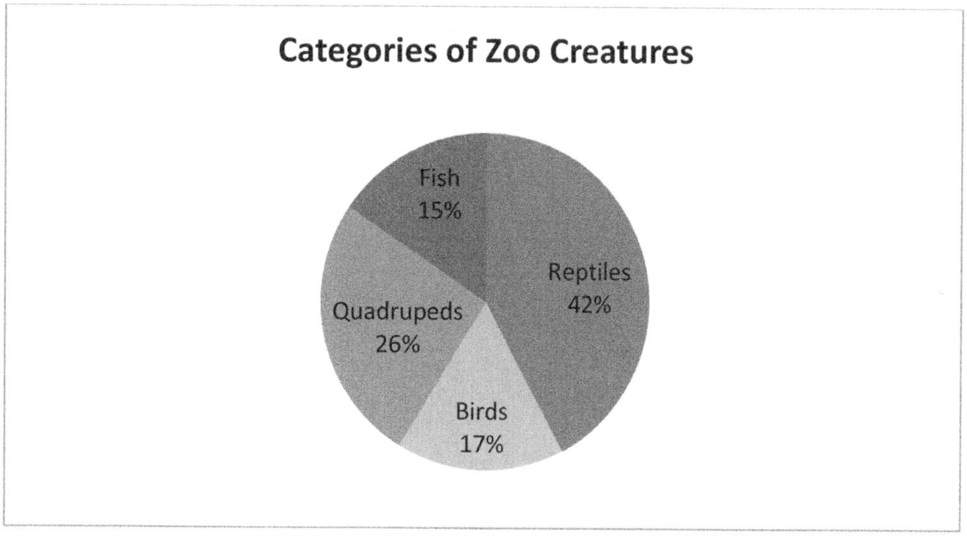

At the end of the year, the zoo still has 1,500 creatures, but reptiles constitute 40%, birds 23%, and quadrupeds 21%. How many more fish were there at the end of the year than at the beginning of the year?

A. 10
B. 11
C. 15
D. 16
E. 150

49. The students at Lyndon High School have been asked about their plans to attend the Homecoming Dance. The chart below shows the responses of each grade level by percentages. Which figure below best approximates the percentage of the total number of students from all four grades who will attend the dance? Note that each grade level has roughly the same number of students.

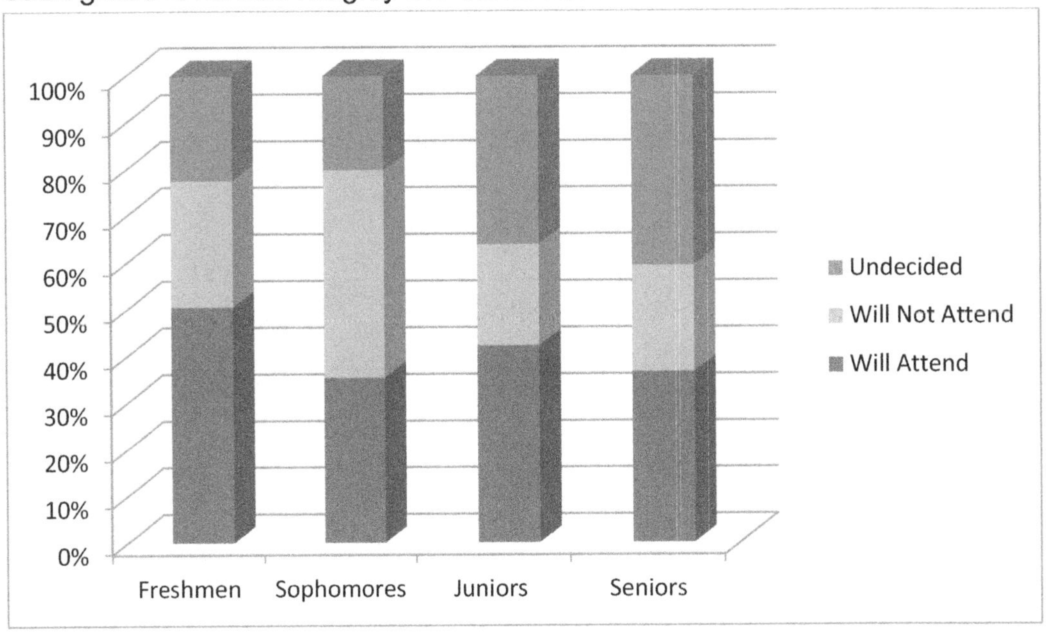

A. 25%
B. 35%
C. 45%
D. 55%
E. 60%

50. Look at the pie chart below and answer the question that follows.

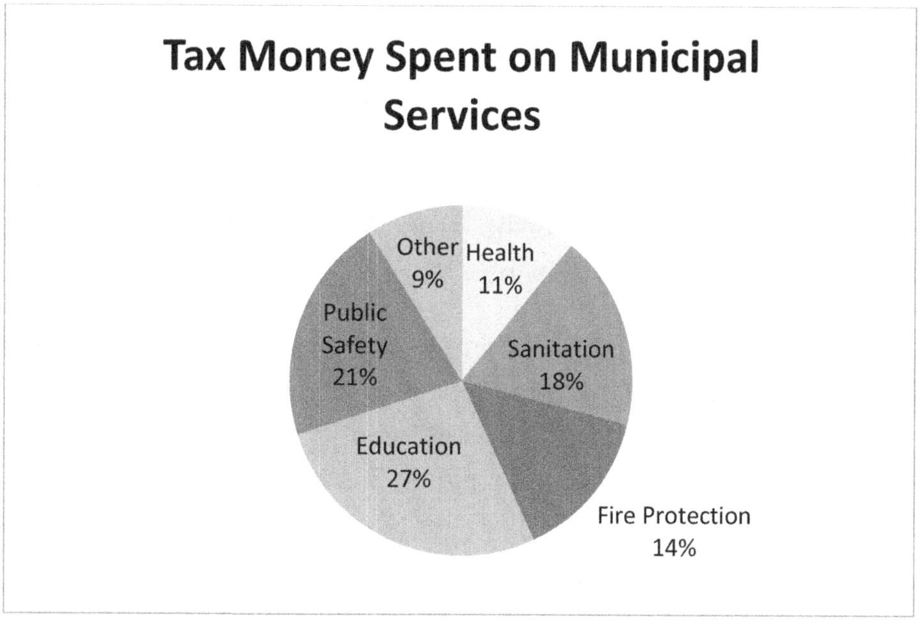

For next year, $6,537,200 in total tax money is budgeted for all municipal services. Each category is allocated the same percentage of next year's budget as the actual percentage spent for the current year. What is the budget amount for public safety?

A. $915,208
B. $1,107,813
C. $1,372,812
D. $1,765,004
E. $1,765,040

CBEST Math Practice Test 4 – Answers

1. The correct answer is A. In order to answer questions on ordering numbers from least to greatest or greatest to least, remember these principles: Negative numbers are less than positive numbers; (b) When two fractions have the same numerator, the fraction with the smaller number in the denominator is the larger According to these principles, $-^1/_4$ is less than $^1/_8$, $^1/_8$ is less than $^1/_6$, and $^1/_6$ is less than 1.

2. The correct answer is B. Perform the division, and then check the decimal places of the numbers. Divide as follows: 1523.48 ÷ 100 = 15.2348. Reading our result from left to right: 1 is in the tens place, 5 is in the ones place, 2 is in the tenths place, 3 is in the hundredths place, 4 is in the thousandths place, and 8 is in the ten-thousandths place.

3. The correct answer is E. STEP 1: Calculate the average high temperature in Fahrenheit. To do so, find the total for all five days and divide the result by 5: (72 + 68 + 65 + 82 + 81) ÷ 5 = 368 ÷ 5 = 73.6°F average. STEP 2: Convert the average in Fahrenheit to Celsius using the formula. To convert Fahrenheit to Celsius, we use this formula: °C = 0.56(°F – 32) = 0.56(73.6° – 32) = 0.56(41.6) = 0.56 × 41.6 = 23.296, which we round to 23°C.

4. The correct answer is B. Divide the total amount by the sales price per unit to solve: $7,375 ÷ $59 = 125 units sold

5. The correct answer is D. First, we can perform division to determine that the plane travels 6.5 miles per minute: 780 miles ÷ 120 minutes = 6.5 miles per minute. Since the plane travels at a constant speed, we multiply this amount by 40 minutes to solve: 6.5 miles per minute × 40 minutes = 260 miles

6. The correct answer is D. STEP 1: Determine the amount of time in seconds: 2 minutes and 48 seconds = (2 minutes × 60 seconds per minute) + 48 seconds = 120 seconds + 48 seconds = 168 seconds. STEP 2: Divide by the number of furlongs to find the rate: 168 seconds ÷ 12 furlongs = 14 seconds per furlong

7. The correct answer is D. The problem uses the phrase '2 out of every 20 employees' so we know that there are 2 employees who form a subset within each group of 20. STEP 1: Take the total number of employees and divide this by 20: 480 ÷ 20 = 24. STEP 2: Take the result from Step 1 and multiply by the amount in the subset to solve: 24 × 2 = 48

8. The correct answer is A. STEP 1: Add the charge for postage and handling to the original price per item: $22 + $3 = $25. STEP 2: Take the result from Step 1 and multiply by the number of items sold: $25 × 32 = $800

9. The correct answer is C. Divide and then express the result as a percentage. STEP 1: Treat the line in the fraction as the division symbol: 6/25 = 6 ÷ 25 = 0.24. STEP 2: To express the result from Step 1 as a percentage, move the decimal point two places to the right and add the percent sign: 0.24 = 24.0%

10. The correct answer is C. Add the four figures together to solve: 163.75 + 107.50 + 91.25 + 10.30 = 372.80

11. The correct answer is B. STEP 1: Find the amount of material needed for each quilt: 2 yards red, 4 yards blue, 1.2 yards gold (6 ÷ 5 = 1.2), and 0.5 yards white = 2 + 4 + 1.2 + 0.5 = 7.7 yards each. STEP 2: Multiply the total number of quilts by the number of yards per quilt to solve: 10 × 7.7 = 77

12. The correct answer is D. Convert the cups to quarter cups: 10 cups ÷ ¼ = 40 quarter cups. Then combine the whole number with the fraction and multiply to solve: $40^1/_4$ × 50 cents per quarter cup = 40.25 × 0.50 = $20.50

13. The correct answer is D. The problem is asking for the total for all five months, so round and then add the amounts together to solve: $700 + $600 + $600 + $800 + $1,000 = $3,700

14. The correct answer is D. STEP 1: Add the whole numbers: 4 + 3 = 7. STEP 2: Add the fractions: 3/8 + 7/8 = 10/8. STEP 3: Simplify the fraction from Step 2: 10/8 = 8/8 + 2/8 = 1 + 2/8 = $1^2/_8$ = $1^1/_4$. STEP 4: Combine the results from Step 1 and Step 3 to solve the problem: 7 + $1^1/_4$ = $8^1/_4$

15. The correct answer is C. In this problem, the fraction on the second number is larger than the fraction on the first number, so we need to convert the first fraction before we start our calculation. STEP 1: Convert $12^7/_{16}$ for subtraction: $12^7/_{16}$ = $11^7/_{16}$ + 1 = $11^7/_{16}$ + $^{16}/_{16}$ = $11^{23}/_{16}$.
STEP 2: There were $8^9/_{16}$ yards left, so subtract the whole numbers: 11 – 8 = 3. STEP 3: Subtract the fractions: 23/16 – 9/16 = 14/16.
STEP 4: Simplify the fraction from Step 3: 14/16 = (14 ÷ 2)/(16 ÷ 2) = 7/8.
STEP 5: Combine the results from Step 2 and Step 4 to get your new mixed number to solve the problem: 3 + 7/8 = $3^7/_8$

16. The correct answer is B. The problem tells us the relative amount of units sold, but the question is asking for the relative amount of units left. So, subtract the decimal from 1 to find the relative amount left: 1 – 0.75 = 0.25. Then multiply the total number of items at the start by this decimal number: 80 items × 0.25 = 80 × 0.25 = 20 items left

17. The correct answer is B. STEP 1: Work out the cost for the usual supplier: 325 pairs × $4 = $1,300. STEP 2: Calculate the price for the second supplier: $1,250 + ($1,250 × .06) = $1,250 + $75 = $1,325.

STEP 3: Compare to the third deal to solve: The third deal is $1,290 so this is the best deal.

18. The correct answer is C. STEP 1: Take the total number of employees and divide this by 5: 250 ÷ 5 = 50. STEP 2: The question asks how many questionnaires have not been completed and returned, so subtract to find the amount in the 'not returned' subset: 5 – 4 = 1. STEP 3: Multiply the result from step 2 by the result from step 1 to solve: 50 × 1 = 50

19. The correct answer is D. STEP 1: Determine the total for sales in December: $20 × 55 = $1,100. STEP 2: Determine the total sales for January: $12 × 20 = $240. STEP 3: Add these two amounts to solve: $1,100 + $240 = $1,340

20. The correct answer is E. He needs to subtract the 92 that he added by mistake to get back to his starting point. Then he needs to subtract 92 again to get the correct result. So, he can subtract 92 two times or simply shortcut by subtracting 184 (92 + 92 = 184).

21. The correct answer is D. The question is asking us about a time duration of 6 minutes, so we need to calculate the amount of seconds in 6 minutes: 6 minutes × 60 seconds per minute = 360 seconds in total. Then divide the total time by the amount of time needed to make one journey: 360 seconds ÷ 45 seconds per journey = 8 journeys. Finally, multiply the number of journeys by the amount of inches per journey in order to get the total inches: 10.5 inches for 1 journey × 8 journeys = 84 inches in total

22. The correct answer is B. For your first step, add the subsets of the ratio together: 6 + 7 = 13. Then divide this into the total: 117 ÷ 13 = 9. Finally, multiply the result from the previous step by the subset of males from the ratio: 6 × 9 = 54 males in the class

23. The correct answer is D. Be careful with the phrase "out of" in proportion questions like this one. We are given the phrase "4 out of 7." If 4 out of 7 prefer the toffee flavor, then the remaining 3 prefer the coffee flavor. Accordingly, the ratio of flavor preference of toffee to coffee is 4 to 3. So, after determining the ratio, your next step is to divide the total by 7: $1540 \div 7 = 220$.Then multiply this by 3 for the coffee-flavor preference: $220 \times 3 = 660$.

24. The correct answer is C. STEP 1: Determine the commission earned per hour: \$15 charged – \$12 paid to employee = \$3 per hour commission. STEP 2: Calculate the total hours that the 10 employees worked: $10 \times 40 = 400$ hours in total. STEP 3: Multiply the total number of hours by the commission per hour to solve: 400 hours × \$3 commission per hour = \$1,200 total commission

25. The correct answer is D. Divide to solve: $49 \div 50 = 0.98 = 98\%$

26. The correct answer is D. Calculate the total, and divide by the number of days. STEP 1: Find the total: \$90 + \$85 + \$85 + \$105 + \$110 = \$475. STEP 2: Divide the result from Step 1 by the number of days: \$475 ÷ 5 = \$95

27. The correct answer is D. STEP 1: Add the whole numbers: $8 + 7 = 15$. STEP 2: Add the fractions: $3/4 + 1/2 = 3/4 + 2/4 = 5/4$. STEP 3: Simplify the fraction from Step 2: $5/4 = 4/4 + 1/4 = 1^1/_4$ = 1 foot and 3 inches. STEP 4: Combine the results from Step 1 and Step 3 to solve the problem: 15 feet + 1 foot and 3 inches = 16 feet and 3 inches

28. The correct answer is C. In this problem, the fraction on the second number is larger than the fraction on the first number, so we need to convert the first fraction before we start our calculation. STEP 1: Convert $28^3/_{10}$ for subtraction: $28^3/_{10} = 27^3/_{10} + 1 = 27^3/_{10} + {}^{10}/_{10} = 27^{13}/_{10}$.

STEP 2: Subtract the whole numbers. $7^9/_{10}$ hours have been spent on the job so far, so subtract the 7 hours: 27 – 7 = 20. STEP 3: Subtract the fractions: 13/10 – 9/10 = 4/10. STEP 4: Simplify the fraction from Step 3: 4/10 = (4 ÷ 2)/(10 ÷ 2) = 2/5. STEP 5: Combine the results from Step 2 and Step 4 to get your new mixed number to solve the problem: 20 + 2/5 = $20^2/_5$

29. The correct answer is C. First of all, convert 96 yards to feet: 36 yards × 3 feet in a yard = 108 feet. Then subtract this from 116 feet: 116 – 108 = 8 feet. Then convert the feet to inches for your answer: 8 feet × 12 inches per foot = 96 inches

30. The correct answer is D. In order to calculate the average, you simply add up the values of all of the items in the set, and then divide by the number of items in the set. To solve problems like this one, set up an equation to calculate the average, using *x* for the unknown value.
Set up your equation to calculate the average, using *x* for the age of the 5th sibling:

$(2 + 5 + 7 + 12 + x) \div 5 = 8$

$(2 + 5 + 7 + 12 + x) \div 5 \times 5 = 8 \times 5$

$(2 + 5 + 7 + 12 + x) = 40$

$26 + x = 40$

$26 - 26 + x = 40 - 26$

$x = 14$

31. The correct answer is D. 3 tablespoons of cocoa powder and ¼ teaspoon of baking powder are needed for the original recipe to make 4 brownies. There are 3 teaspoons in a tablespoon, so calculate the total teaspoons needed for the original recipe first: 3 tablespoons × 3 = 9 teaspoons cocoa powder + ¼ teaspoon baking powder = 9¼ teaspoons in total. We are now making 12 brownies, so we need to multiply all of the ingredients by 3: 9¼ × 3 = 27¾ teaspoons

32. The correct answer is B. First add up all of the numbers to get the total: 35 + 42 + 23 + 20 = 120. We divide the number of office workers who live in the country, which is 20 people, into this total to get our answer. We can find this amount in the lower right-hand corner of the chart. So, set up the fraction and then simplify: 20 ÷ 120 = $^{20}/_{120}$ = $^{2}/_{12}$ = $^{1}/_{6}$

33. The correct answer is B. First, we will use variable *T* as the total number of items in the set. The probability of getting a red scarf is $^{1}/_{3}$. So, set up an equation to find the total items in the data set.

$$\frac{5}{T} = \frac{1}{3}$$

$$\frac{5}{T} \times 3 = \frac{1}{3} \times 3$$

$$\frac{5}{T} \times 3 = 1$$

$$\frac{15}{T} = 1$$

$$\frac{15}{T} \times T = 1 \times T$$

$$15 = T$$

We have 5 red scarves, 6 blue scarves, and *x* green scarves in the data set that make up the total sample space, so now subtract the amount of red and blue scarves from the total in order to determine the number of green scarves.

$$5 + 6 + x = 15$$

$$11 + x = 15$$

$$11 - 11 + x = 15 - 11$$

$$x = 4$$

34. The correct answer is C. Isolate the integers to one side of the equation.

$\frac{3}{4}x - 2 = 4$

$\frac{3}{4}x - 2 + 2 = 4 + 2$

$\frac{3}{4}x = 6$

Then get rid of the fraction by multiplying both sides by the denominator.

$\frac{3}{4}x \times 4 = 6 \times 4$

$3x = 24$

Then divide to solve the problem.

$3x \div 3 = 24 \div 3$

$x = 8$

35. The correct answer is A. Remember that to solve problems like this, you need to deal with the integers (whole numbers) and then isolate the variable (x). The solution is as follows:

$4x - 3 = 2$
$4x - 3 + 3 = 2 + 3$ (Add 3 to both sides of the equation.)
$4x = 5$
$4x \div 4 = 5 \div 4$ (Divide both sides by 4.)
$x = \frac{5}{4}$

36. The correct answer is A. The total amount that Toby has to pay is represented by C. He is paying D dollars immediately, so we can determine the remaining amount that he owes by deducting his down payment from the total. So, the remaining amount owing is represented by

the equation: C – D. We have to divide the remaining amount owing by the number of months to get the monthly payment (P): P = (C – D) ÷ M = $\frac{C-D}{M}$

37. The correct answer is B. We need to set up a fraction, the numerator of which consists of the amount of sales in dollars for sweatpants, and the denominator of which consists of the total amount of sales in dollars for both items. The problem tells us that the amount of sales in dollars for sweatpants is 30*s* and the total amount of sales is 850, so the answer is 30*s*/850.

38. The correct answer is A. The equation is: F = $500P + $3,700. We are told that the total funds are $40,000 so put that in the equation to solve the problem.

$40,000 = $500P + $3,700

$40,000 – $3,700 = $500P

$36,300 = $500P

$36,300 ÷ 500 = $500 ÷ 500P

$36,300 ÷ 500 = 72.6

39. The correct answer is C. Add 9 to each side to get rid of the integer on the left.

$3x - 9 = -18$

$3x - 9 + 9 = -18 + 9$

$3x = -9$

Then divide each side by 3 to solve.

$3x \div 3 = -9 \div 3$

$x = -3$

40. The correct answer is C. If door A is locked with the red key, then door B is locked with the blue key. If the key that locks door B also locks door D, then the blue key is also used to lock door D.

41. The correct answer is C. Since we are dealing with a square, all four sides of the floor are equal to each other. The tiles are also square, so they also have equal sides. Therefore, we can simply divide to get the answer: $64 \div 4 = 16$

42. The correct answer is C. To solve problems like this one, try to visualize two rectangles. The first rectangle would measure 8×9 and the second rectangle would measure $X \times Y$. Essentially a rectangle is missing at the upper left-hand corner of the figure. We would need to know both the length and width of the "missing" rectangle in order to calculate the area of our figure. So, we need to know both X and Y in order to solve the problem.

43. The correct answer is C. Calculate the total square footage of the room first: $12 \times 16 = 192$ total square feet. Then calculate the area under each bookcase. Bookcase A: $9 \times 1.5 = 13.5$; Bookcase B: $7 \times 1.5 = 10.5$; Total bookcase area: $13.5 + 10.5 – 24$ square feet. Next find the area of the room without the bookcase area: $192 – 24 = 168$ square feet to be covered. Finally, determine the cost: 168 square feet × \$4.25 per piece = \$714 total cost

44. The correct answer is A. The area of a rectangle is equal to its length times its width. This football field is 30 yards wide and 100 yards long, so we can substitute the values into the appropriate formula.
rectangle area = width × length
rectangle area = 30 × 100
rectangle area = 3000

45. The correct answer is B. You are being asked about the distance around the outside, so you need to calculate the perimeter, Write out the

formula: (length × 2) + (width × 2). Then substitute the values: (5 × 2) + (3 × 2) = 10 + 6 = 16

46. The correct answer is C. For ten hamburgers, the total price is $85, so each hamburger sells for $8.50: $85 total sales in dollars ÷ 10 hamburgers sold = $8.50 each

47. The correct answer is C. The quantity of diseases is indicated on the bottom of the graph, while the number of children is indicated on the left side of the graph. To determine the number of children that have been vaccinated against three or more diseases, we need to add the amounts represented by the bars for 3, 4, and 5 diseases: 30 + 20 + 10 = 60 children

48. The correct answer is C. At the beginning of the year, 15% of the 1,500 creatures were fish, so there were 225 fish at the beginning of the year (1,500 × 0.15 = 225). In order to find the percentage of fish at the end of the year, we first need to add up the percentages for the other animals: 40% + 23% + 21% = 84%. Then subtract this amount from 100% to get the remaining percentage for the fish: 100% – 84% = 16%. Multiply the percentage by the total to get the number of fish at the end of the year: 1,500 × 0.16 = 240. Then subtract the beginning of the year from the end of the year to calculate the increase in the number of fish: 240 – 225 = 15

49. The correct answer is B. The dark gray part at the bottom of each bar represents those students who will attend the dance. 45% of the freshman, 30% of the sophomores, 38% of the juniors, and 30% of the seniors will attend. Calculating the average, we get the overall percentage for all four grades: (45 + 30 + 38 + 30) ÷ 4 = 35.75%. 35% is the closest answer to 35.75%, so it best approximates our result.

50. The correct answer is C. Take the total dollar amount of the budget and multiply by the 21% for public safety: $6,537,200 × 0.21 = $1,372,812

Made in the USA
Coppell, TX
10 April 2022

76334713R00109